THE QUANTUM HERMETICA

A Documentation of the Parallels between
Hermetic Occult Science and Modern Physics

By Olivia & Ethan Palmer
Foreword by Joseph Patterson

Table of Contents

Foreword..4
I Introduction..6
II The Principle of Mentalism....................................11
III The Principle of Correspondence.........................24
IV The Phenomena of Mentalism and Correspondence.........35
V The Principle of Vibration....................................52
VI The Phenomena of Vibration...............................57
VII The Principle of Polarity....................................66
VIII The Principle of Rhythm...................................70
IX The Principle of Cause and Effect......................73
X The Principle of Gender......................................79
XI Epilogue..86
References..95

Foreword

When I first met Olivia and Ethan back in college in the mid-nineties they were an eccentric couple. It was not until I got to know them better that I soon realized just how eccentric they were. Ethan was studying electrical engineering and Olivia was a philosophy major. However, what they shared in common that at first I never got, was a penchant for the esoteric. Among esoteric subjects they had a special interest in Hermeticism.

At the time, I could not believe that such interests were compatible with their otherwise serious intellectual life. Yet there they were. One night at a bowling alley the topic came up. Ethan told me he believed these things to be real based on various purportedly mystical experiences he had had, but moreover that one day he expected such things to have scientific explanations.

I scoffed at this since at the time my mind was in a *materialistic* way of thinking about the world. Such things could not be scientific I insisted, as they are incompatible with the physicalist outlook upon which I believed science was built. Ethan insisted otherwise, and brought up quantum physics as an exception. Of course, in my youthful naiveté, I scoffed at this possibility.

Nevertheless, whenever I would have subsequent discussions with them on this topic, they seemed calmly intellectual about it. It was not the approach you would expect from a couple interested in what I had conceived of at the time as starry-eyed mysticism.

Since college, I would continue to have various correspondences with them over the years. Sometimes these topics came up again. Whenever they did, I was struck by the academic tone of the conversations. They had arguments and evidence, and I had a tied tongue.

Despite this I still stuck to my materialistic way of interpreting the world, since I felt it was the "status quo" of looking at how the universe works. Any resistance to the status quo in my mind really only exposed mystical or religious motivations and were beyond the confines of reason. What they brought up I felt were really only gaps in our understanding. One day in the future I believed scientists would discover a *materialist explanation* of these gaps of knowledge, thereby relegating these immaterial anomalies to the same trash pile as religion and mythology. In my mind such anomalies only represented current gaps in our knowledge of science, and were akin to saying "God did it."

However gradually, I began to realize that the reason I was dismissing them was due to a conflation in my mind between science and my scientific materialism. My motivations were not based on a careful examination of the evidence or trying to better explain the evidence, but rather upon a metaphysical prejudice to dismiss evidence that did not fit my physicalist view of the world.

What they were offering by contrast, was not incompatible with science in any way, nor did it demonstrate any gaps in a scientific picture. It was surprisingly wholly scientific, a view of the world based on the scientific immaterialism of John Archibald Wheeler. Furthermore, it had far more explanatory power and parsimony than the contrasting physicalist view I had held for many years. On the grounds of the science presented in this book alone one might argue that there are materialistic explanations that are possible for them, but they are not at all probable. There is a difference between possibility and probability, and in this case that difference is vast.

Before one goes to make the argument that non-materialistic interpretations will always be pseudoscience since anything non-material is outside science, I want to make one thing clear to the readers. I know how science works and I know that there are certain groups of people that deny scientific facts by saying "it's only one interpretation of the evidence." Good examples of these are the flat-earthers or the creationists. They may say that views opposing theirs have the same evidence but different interpretations, and in a trivial sense that may be true. However, the evidence directly contradicts these interpretive frameworks themselves. Furthermore, the framework of a creationist or a flat-earther is ad hoc, and sometimes those interpretations have to deny entire fields of science. This is how pseudoscience is done, and regretfully I recognized that my view of the world before was resembling exactly such a collection of ad hoc excuses.

To make my point more obvious you can create any interpretation you want in science. In science you have a big sheet of data points as Popper pointed out, and you can connect those data points however way you like and make it "fit" any framework you want, even if the framework is as ridiculous as flat earth theory. The issue is if you do this, then science will not work, and you can literally change the laws of physics however way you want in order for it to "fit" any beliefs you might have about the world. The true goal of science is to find the *best explanation* of the evidence, not to make different "interpretations" to fit any particular metaphysical framework. I realized that I and many others had been and continue to do exactly this though when it comes to a physicalist view of the world.

Once I recognized this, I discovered that it opened up a whole new way of thinking about the world. Things that I would previously have dismissed became possibilities to me. Then one night a few years ago, Ethan called me up on the phone telling me he had done it. He had cracked Hermeticism with new discoveries in modern physics. Moreover, he told me that I would be astonished at just how simple it was. Olivia meanwhile had a metaphysical model that fit the same essential picture and was equally simple. I was skeptical at first, but as Ethan explained it, I became more convinced. It had been hiding right under everyone's collective nose the whole time.

When they later told me they had written a book on it I felt a little nervous. Then I realized the knowledge in this book needs to be shown to the public. For far too long has this information been hidden from the world, and it is time for us to discover the true nature of reality.

For readers that are open minded about new ideas I encourage you to read this book from its first chapter to the last and to understand its implications. There will be times that you may have a gut feeling that some of this stuff can't possibly be true because it is too weird and defies common sense. However as with all things in science, reality does not care about intuition or feelings, and it may be the case that reality truly is stranger fiction. A reality that is in fact not material at bottom, but is completely and fundamentally mental. Hopefully this book will change the way you believe and perceive reality, and perhaps one day everyone can have this knowledge of the world as well.

-Joseph Patterson

I Introduction

Hermeticism is a metaphysical system forming the basis for many esoteric and occult schools of thought, reputed to date back to ancient Egypt. According to legend, the Hermetic ideas first originated with Hermes Trismegistus, or Hermes the Thrice Great, whom the Egyptians identified with their god of wisdom, Thoth, the Greeks with Hermes, and the Romans with Mercury. These ideas were said to have been written down in a vast body of literature, called the Corpus Hermeticum.

Much was lost with the burning of the Library of Alexandria. However, forty-two Hermetic texts survived through the dark ages and into modern times. A revival of Hermeticism took place during the Renaissance, when these old texts were reexamined again. At this time, it was believed that these originated in Egypt's ancient past, and that Hermes Trismegistus was a real historical personage.

Then in 1614 they were reexamined by Swiss philologist, Isaac Casaubon. Casaubon's conclusion was that the texts dated to the 2nd century AD based on their linguistic style, which was compatible with Greek style of that era rather than that of an ancient Egyptian priest.

It was not until later that the flaw in Casaubon's methodology was discovered by Ralph Cudworth. Just because the writing style was Greek did not mean the ideas originated from Greek thinkers, and in fact it was likely that they merely wrote down earlier Egyptian thought.[1] Additionally, the ancients had a different conception of authorship than we do in the modern era. In the ancient world authorship was thought to concern the origination of the ideas put down in the text rather than the writer of the particular manuscript. Additionally, Egyptian elements have since been demonstrated to exist in the Hermetic texts to a greater degree than previously thought.[2]

Given the parallels drawn between Hermetic philosophy and the religious concepts of ancient Egypt, a more plausible hypothesis presents itself. It is entirely likely that with the Hellenization of Egypt following the time of Alexander the Great, wisdom teachings passed down from the Egyptian mystery schools were studied and rewritten in newer texts by Greek scholars. With the burning of the Library of Alexandria around the same time, and much of the written knowledge of Egypt being stored there, the idea that the current Hermetic texts are Greek copies of the original material seems to have merit.

This being the case it is indeed possible that the Hermetic teachings were passed down from deep antiquity as legend would have it. And if so, who can tell how much legend surrounding the Hermetica blurs into fact? Perhaps we may never know the full story.

There is however another remarkable and mysterious clue to which this book is dedicated, which would seemingly vindicate the idea that the Hermetic teachings are more than just a repackaging of Greek philosophy. A newer Hermetic document was published in 1908 by the Masonic Lodge of Chicago called *The Kybalion*. The origins of some of the ideas contained in it of course remain mysterious, as it was produced by a secret society to which public access is restricted. However, it faithfully summarized the ancient Hermetic philosophy in terms of seven fundamental principles in a more compact form.

The remarkable part is that the concepts contained therein have precise parallels to modern and in some cases cutting edge physics. The caveat is that many of these parallels will only become apparent when the physics is cast within an ontology of mind rather than that of the materialism naively assumed throughout much of modern science.

The tie in with mental metaphysics will not be based merely on abstract philosophic consideration of the mind either. Though philosophic consideration of these topics will become relevant as well, the Hermetic principles will be described with hard and fast scientific models corroborated with evidence, and having foundations in mathematically precise theories. As such, attempts at dodging the conclusions with facile instrumentalism or specious attempts at referring to them as "woo," will be exposed as a naked means of avoiding the conclusions without argument. Here the reader is encouraged to remember that criticisms given without argument or evidence can equally be dismissed without argument or evidence.

As these principles are accounted for with physics, we will also discuss relevant phenomena showing how a wide variety of occult, paranormal and even religious concepts can also be readily explained within their framework. In so doing none of these will appear mysterious or mystical. Occult concepts will no longer be seen as beyond the domain of science or reason and incomprehensible to the mind of man. Rather the spiritual domain will be shown to be explicable with the sciences, or as *The Kybalion* would put it:

"The Principles of Truth are Seven; he who knows these, understandingly, possesses the Magic Key before whose touch all the Doors of the Temple fly open." --The Kybalion

In deriving them it will be shown that the seven principles are connected with one another such that they build on one another. The aim here is to assist the reader in placing the Hermetic sciences into a cohesive framework from which he or she may comprehend the whole. The seven principles and the essential science behind them are as follows.

1. The Principle of Mentalism: This principle tells us that "All is mind, the universe is mental." This means that the only substance in existence is mind, and that what we mistakenly call the "physical" universe, is nothing more than an immaterial construction within a Universal Mind, called God by religion, The Mind of The All by Hermeticists, or the Larger Consciousness System by some physicists.[3] To derive this principle we will begin by surveying the extraordinary findings of recent years in quantum mechanics falsifying physical realism. This will be followed up by the even more extraordinary discoveries in quantum gravity demonstrating that the fabric of space-time itself is nothing more than an illusory construction generated by quantum information existing beyond physical reality. Here the direct parallel with a computer game emerging from the information in its program will become apparent.

The ontological implications of this will then be considered. Once they are contrasted with the facts of conscious experience as shown both from philosophy of mind, but more readily from the reader's immediate and ineffable first person awareness of the facts of his own mind, the conclusion of mentalism will be obvious.

All that will be needed then is a physics based description of consciousness compatible with these facts, of which cognitive science is already helping us. Such theories include the Conscious Realism of Donald Hoffman, as well as the field of quantum cognition at large. When combining these with the physics of emergent space-time, a fully scientific account of the Principle of Mentalism will be demonstrated, revealing space-time to emerge from cognitive quantum information. Lastly, this will be shown to be compatible with the facts of neuroscience through quantum biology.

2. The Principle of Correspondence: Once mentalism is understood the second most important principle of Hermeticism will be derived from it; The Principle of Correspondence. The Principle of Correspondence as summarized in Hermeticism is "As Above, So Below" or alternately "As Within, So

Without." The Principle of Correspondence relates the higher spiritual world to the lower informational construction we call the physical world within it.

The idea here is that the higher world above corresponds to the lower world below such that phenomena going on in the spiritual world manifest in patterns in the physical world. In other words, the activities of the computer code in the harddrive correspond to the goings on in the computer game. Thus the two worlds mirror each other.

The "As Within, So Without" formulation of this principle can readily be understood with the combination of the concept of quantum cognition and the science of emergent space-time. Here it is shown that the realm beneath space-time is no mystery at all, but rather that the individual already has direct access to it, as his or her introspective consciousness is already a natural part of this realm.

Just as the physics behind the Principle of Mentalism demonstrates a philosophy of mentalism, so too the physics behind the Principle of Correspondence demonstrates Platonism. The quantum information residing within the quantum mechanical wave-function is identical with the Platonic forms and archetypes of the Platonic world that correspond to their instantiations in the physical world.

3. The Principle of Vibration: This principle states that "Nothing rests; everything moves; everything vibrates." According to the Principle of Vibration everything in the universe is comprised of vibrating energy. Everything we see as a material object is actually the result of these vibrations. This is not limited to matter and energy either, but even to space-time itself, which *The Kybalion* refers to as the ether.

The parallels with quantum mechanics here are obvious. In quantum theory every physical object possesses a wave-particle duality, and exists as a literal vibrating wave of probabilities while not observed. More interestingly as will be explored in the chapter on vibration, more advanced physics such as superstring theory, as well as other ideas in quantum gravity, are confirming what *The Kybalion* stated about space-time or "the ether" being "matter at a higher degree of vibration." Likewise, as we shall see, the vibration of The All being stated to be "at rest" in *The Kybalion,* exactly matches what modern physics tells us about the state of the Universal Wave-function.

4. The Principle of Polarity: The Principle of Polarity states that "Everything is dual, everything has poles; everything has its pair of opposites; opposites are identical in nature, but different in degree." When we examine the details of this principle as described in *The Kybalion* we find that it is the product of the Principle of Mentalism in combination with the Principle of Vibration. All is mind, and thus what we refer to as material objects are actually mental states. Likewise, these mental states change as their accompanying vibration changes.

The science behind this readily follows from the science behind these other two principles. Specifically, the science of quantum cognition relating conscious states to quantum probability waves as seen in Conscious Realism demonstrates such a relation between change in conscious state and change in vibratory frequency. We will then examine a direct example of polarity within the physical sciences in the quantum biological phenomena of quantum smell.

5. The Principle of Rhythm: The Principle of Rhythm states that "Everything flows out and in; everything has its tides; all things rise and fall; the pendulum swing manifests in everything." Here the Principle of Rhythm can be easily seen to derive from the Principle of Vibration. Rhythm is after all built directly into vibration as its periodicity. In turn given that everything is built up from vibrating waves of probability, as quantum mechanics reveals, rhythms will naturally build up at all scales from Fourier series of these waves. Given that quantum cognition treats these vibrations as being mental in nature, just as Hermeticism reveals, these rhythmic patterns can be seen in the phenomena of mind as well. This includes everything from mood swings, all the way up to long term trends of civilizations over periods of time.

6. The Principle of Cause and Effect: The Principle of Cause and Effect is a direct consequence of the Principle of Correspondence. As *The Kybalion* tells us; "Every Cause has its Effect; every Effect has its Cause; everything happens according to Law." This is of course intuitively obvious. However, the sort of cause and effect it is speaking of is not so much the mundane cause and effect of physical determinism. Rather it is speaking of causes on one plane inducing effects on other planes. Just as the Principle of Correspondence correlates the activities of planes, so too the Principle of Cause and Effect points out that some of these correlations can include causations from one plane to another as well.

An examination of modern parapsychological data will be explored here, showing that this is indeed the case. In turn this will be described with the same physics used to model Correspondence. Quantum cognitive causes in the plane behind space-time, described as Hilbert Space with modern mathematics, have effects on the plane of physical space-time that emerges from it.

It should be noted that the Principle of Cause and Effect also says that "Chance is but a name for Law not recognized." Here the astute reader may wonder how this could be compatible with the fundamental quantum indeterminacy seen in nature. On a closer examination of the physics of consciousness, it will be shown that this is identical with what is referred to as free-will, and that this does not so much escape the principle as it grounds it. All chains of cause and effect must have first causes, and in a mental universe, Mind is its first cause. Likewise these are the smaller first causes of smaller conscious agents, albeit constrained by the range of probability given by the wave-functions which describe them.

7. The Principle of Gender: This principle tells us that: "Gender is in everything; everything has its Masculine and Feminine Principles; Gender manifests on all planes." At first, this principle seems unrelated to the sort of physics described above, but on closer examination it will be shown to derive quite naturally from Mentalism. In turn everything described by Mentalism can equally well be described with the mentalistic physics from before, such as Hoffman's Conscious Realism.

Here gender is not explicitly referring to biological sex. Though as will be shown, biological sex is but one manifestation of it. Rather what it refers to is a binary structure that manifests as a direct consequence of the process of generation or creation within the mental universe.

If we start only with The Mind of The All we find that this Mind must create within itself. What it creates though is also a mind. This parallels with the conscious agent dynamics described by the cognitive scientist Donald Hoffman. Here conscious agents can be entangled with one another, forming a single mind or quantum information system greater than the sum of the individual conscious agents. Likewise, they can be decomposed from one another, to form smaller conscious agents. And if The Mind of The All wishes to create, the process of creation can be described in this manner.

When someone creates, they create what they love to create. A poem for instance is not created by someone who dislikes poetry. Thus the origin of the gender binary can be seen here. A lover initiates the creation of what is loved, and the loved receives the love of the lover. These principles of initiating and receiving love are called masculine and feminine respectively by Hermeticists. As will be shown, this is paralleled in our common every day understandings of male and female as well. Though of course this does not mean that only males create, or that only females receive love, as masculine and feminine principles are seen in everything.

Additionally, this forms the basis for gender complementarity, as something is naturally complementary to what it is cut out from. In turn this pattern of female being cut out of or individuated from male, can be seen in both the Hermetic teachings of *The Kybalion* as well as a vast number of religious doctrines on the matter spread throughout both geography and time. As will be shown though, none of this needs to be grounded in blind faith or religious dogma. Rather all of it reduces to the Hermetic Principle of Gender, which in turn can be fully described by the dynamics of conscious agents in the physics of consciousness.

Note To The Reader: As the book progresses, each of these principles and their associated phenomena will be looked at in more depth. Parallels to parts of *The Kybalion* will be shown in italics for comparison to better benefit the reader in recognizing the connections between the physics on the one hand and the occult science on the other.

As will become quickly apparent in the reading of this book, the concepts in this book are firmly supported by established science yet also overlap into many concepts directly associated both with what has been called the supernatural, as well as with religion. However, this is not a religious text! Everything here can be confirmed independently by the reader and are facts available to members of all religions and no religion at all.

Our hope is that by the end of this book the reader will have an understanding of a great many phenomena previously thought to be mysterious and incomprehensible, dismissed as superstition, or accepted only as a matter of faith. By placing the entire Hermetic philosophy within a completely scientific framework it is hoped that this ignorance can be alleviated. Furthermore, we shall show how the essential framework can at the same time be immediately known from facts readily available in the reader's own conscious experience. In this the reader will be able to have an immediate awareness of the reality in which he or she exists, rather than merely an abstract understanding of these principles. It is in this spirit of knowledge rather than ignorance that the book is written. With this the reader is left to enjoy the adventure ahead.

II The Principle of Mentalism

The Principle of Mentalism is the simplest and yet the most important of the Hermetic principles. From this one principle the general framework of Hermeticism can be derived. This principle tells us in simple terms that:

"The universe is mental, held in The Mind of The All."

Though being the simplest of the principles, it is also perhaps the most counterintuitive. After all, the world around us appears to be quite material. In fact, the very notion that it could be otherwise sounds like nonsense at first glance.

Moreover, there is the problem of how such a principle could possibly be translated into science. Science is thought to be about the material world to such an extent that in many people's minds the terms "science" and "materialism" are synonymous. Despite the common beliefs, on closer inspection of the fundamental physics describing the nature of "matter," this is not the case. It is found that "real matter" does not exist at all!

Thus to begin to understand how science can arrive at Mentalism, we must start by looking at this fundamental physics. In fact, the conclusion of the discoverer of this physics concluded as a matter of his science, not his philosophy, that the Principle of Mentalism is true:

"As a man who has devoted his whole life to the most clearheaded science, to the study of matter, I can tell you as a result of my research about the atoms this much: There is no matter as such! All matter originates and exists only by virtue of a force which brings the particles of an atom to vibration and holds this most minute solar system of the atom together... We must assume behind this force the existence of a conscious and intelligent Spirit. This Spirit is the matrix of all matter."[1] *~Max Planck*

As we proceed through this chapter, we shall soon see why Planck's scientific conclusion is quite unavoidable.

A. Quantum Mechanics

The first clue to the unreality of matter arises in the study of quantum mechanics. To understand quantum mechanics, and how it leads necessarily to the conclusion of immaterialism, it is necessary to start with a little of the history of quantum mechanics. Along the way certain key experiments will become relevant.

i. The Origin of Quantum Theory: Quantum theory began in 1900 with the discovery by Max Planck that a problem associated with blackbody radiation could not be solved unless light was treated as a particle. This particle was called a photon and had a discrete amount of energy called a quantum. The confusion lay around the fact that Thomas Young had shown light to be a wave in the double slit experiment almost a century earlier in 1801.

ii. The Double Slit Experiment: Young's experiment demonstrated that light sent through a pair of slits in a barrier radiated equally from those two slits in a series of concentric semi-circular waves. As those waves crossed paths they would interfere with each other, forming a series of bands on a film placed behind the slits. If light were a particle however it was thought that those series of bands would not exist, as no concentric wave-fronts would exist to create them. Rather if they were particles, they would simply form a pair of clumps on the film where the photons were projected through the slits, like bullets from a rifle.

Despite Young's experiment showing light to be waves, Planck demonstrated that light had to be comprised of particles. Thus light behaved both as a wave and yet like a particle, which seemed paradoxical. To make things more complicated, in 1924 Louis de Broglie demonstrated that just as waves of light are somehow particles, elementary particles of matter such as electrons and protons behave as waves which form interference patterns. This state of affairs was extraordinarily confusing, as these two facts seemed to be contradictory to one another.

To resolve this, Erwin Schrodinger proposed a strange idea in 1926 with the discovery of the Schrodinger wave equation. His solution was to suggest that the photon and the electron are indeed both particles and waves. When they are observed, they are particles with defined states and locations in space. When they are not, they do not have defined states but are smeared out in a wave of probable locations and states all in superposition with one another. The mathematical description of this wave of probability is called the *wave-function* and is arrived at by solving Schrodinger's equation.

This does not make sense with our natural intuitions about the world. Matter as we conceive of it has a location in space and well-defined properties whether or not we are looking at it. But the physics tells us it does not.

So revised experiments were conducted with single particles being sent through the slits one at a time to see if perhaps the particles were interacting with each other to produce a wave pattern. With only one particle going through at once, there is literally nothing else for it to interfere with. Thus it ought to behave as a particle.

When the experiment was conducted however, the individual particles once again collectively formed an interference pattern on the screen. Clearly they had to be interfering with something, and yet there was nothing else other than themselves to be interfering with. Thus astonishingly, it was found that they really are spread out in a probabilistic wave of possible positions, rather than actual ones as they went through the slits. In turn, this probability wave interferes with itself by passing through both slits of barrier at the same time.

iii. The Observer Effect: Now if one were to attempt to measure or observe which of the two slits the electron actually goes through as it is going through one or the other, it is suddenly found that the electrons produce clump patterns once again rather than wave patterns. Merely observing the electron caused it to take on a definite state as a particle with defined properties. This effect of taking on definite states upon observation is called the *"collapse of the wave-function."*

Of course here the question will arise as to what constitutes an "observation." Many will say that the observation is in fact only the measurement of an experimental apparatus, or that it is merely the interaction with any other physical object. They would not at all be wrong in asserting this.

However, they *would* be wrong in asserting that this means that consciousness does not also participate in measurement. After all, experimental apparatuses are themselves made of quantum particles that must themselves also be measured to take on definite states. So who or what collapses their wave-functions? We quickly see here that collapse is not some objective event that happens only with respect to one object or another. Rather it is subject to the reference frame of who or what is doing the measurement. And this necessarily includes the conscious mind as one of these reference frames.

Some will counter that decoherence collapses the wave-function long before any conscious mind observes it. The question arises then as to what causes this decoherence. Advocates of decoherence will say that it is the environment.[2] This answer is insufficient for two reasons. Firstly, the environment is itself made of more quantum particles. So why would the collective wave-function of particles comprising the environment be collapsed if someone or something has not observed them yet as well? Secondly why isn't consciousness part of the environment? On examining these questions, it becomes obvious that the assumptions underlying decoherence can not be consistently applied.

The sorts of materialists who make these claims also reject the notion that the mind is anything beyond science. So if they do not see the mind as special, they would have no grounds on which to specially *exclude* the mind from collapsing the wave-function if *every other object studied by science does*. In fact, the only way in which consciousness can be excluded as a potential quantum measurer is through special pleading. To do this the mind must be made an exception to the physical world, exactly as it is in the same dualistic doctrines these materialists at the same time oppose. This betrays an unjustified prejudice posited solely for the purposes of attempting to justify materialism against of the weight of the scientific evidence.

It is at this point that we begin to see the connection between the fundamental level of physics and the mentality of the world. Fundamental physics is affected in some way by the mind through the observer effect. However we still have a ways to go before we can actually establish that the only substance is mind, and that what we call the physical universe is but a construction within Mind.

iv. Physical Realism and Hidden Variables: Einstein famously did not like implications of quantum mechanics. He found this notion of physical properties not existing before you look to be incompatible with the very notion of science.

To make his point, he conceived of a thought experiment with Boris Podolsky and Nathan Rosen wherein two particles, one in one state, and another in a complementary opposite state, were placed in superposition together in the wave-function.[3] If these were separated at a vast distance, then the very act of measuring one of the particles would determine its state, and thereby also instantly determine the complementary state of the other particle many miles away. The idea that one particle could cause effects on another miles away with no physical mechanism in between to account for them he referred to as "spooky-action-at-a-distance." This "spooky-action-at-a-distance" seemed absurd to him though, and incompatible with a scientific view of the world.

In Einstein's mind quantum mechanics could not be the full picture. He believed there had to be some hidden variable present in the physics that was causing the particle to have a real defined physical state before it was measured. If these hidden variables existed, he reasoned they would have to be "local," meaning they would have to be located in the same spatial vicinity of the particle so as to be able to affect it. This belief in local hidden variables is what is called *local realism.*

Later another physicist, John Bell, proposed a mathematical inequality that could be used to test Einstein's thought experiment. Alain Aspect put this inequality to the test in 1981, demonstrating that this spooky-action-at-a-distance was indeed a reality.[4] This is of course what is known as quantum entanglement in many popular accounts of the subject. If there exist any hidden variables, they are not local to the vicinity of the particle being affected, meaning local realism is false.

This posed a problem. The collapse of the wave-function occurs instantly. So if there exist any hidden variables giving the pre-existing quantum state a physical reality, they must be influencing the second particle infinitely fast. Yet according to special relativity nothing can travel faster than light, including any hidden variables. Thus non-local realism also violates the laws of physics and is impossible.

This fact alone demonstrates that whatever is linking these entangled particles can not be physical, as physicality entails existence within space and time. So whatever links these particles can not be in space-time, as Einstein's relativity tells us.

To back this up, another experiment was conducted in 2007 by the German physicist Anton Zeilinger.[5] This was a test of another inequality proposed by Anthony Leggett to test for non-local realism. As expected this test disproved non-local realism as well. With both local and non-local realism being proven to be false, there are no other possible categories in which physical realism could be true. As Zeilinger's concluded, physical realism must be ruled out. Therefore, what we naively call matter does not exist. Rather only the appearance of matter exists.

Some say that these results apply only to the microscopic world, and do not apply to the everyday world of our experiences. However, there is nothing in science that restricts one set of laws to one level of reality and another set to another level of reality. Instead the true laws of physics are uniform at all levels. What are said to be laws applicable only at some levels are in actuality only approximations of the ultimate laws at a specific level. Thus while we may not be aware of the effects of quantum mechanics in our experience, they still apply to the realm of our experience.

The notion that the laws of quantum mechanics can be excluded from the realm of everyday life is the hypothesis of macrorealism. A third inequality, known as the Leggett-Garg inequality, has been used to test this hypothesis as well. Macrorealism failed these experiments,[6] just as realism has the others.

Since then many other experiments have been conducted proving the falsity of realism, which the reader may explore in his or her own time. Among these are the famed quantum eraser experiments,[7] the test of the Kochen-Specker theorem,[8] and the bizarre quantum Cheshire Cat experiment, in which a particle has been shown to be reducible to disembodied immaterial properties.[9] All of these experiments show that the concept of "scientific materialism" is simply erroneous and based on ignorance of the actual science involved. As one of quantum mechanics' other founding fathers, Werner Heisenberg,[10] noted years ago, notions of materialism are strictly pseudoscientific:

"The ontology of materialism rested upon the illusion that the kind of existence, the direct "actuality" of the world around us, can be extrapolated into the atomic range. This extrapolation is impossible, however." –Werner Heisenberg

Some have argued that no one really understands quantum mechanics and thus that no conclusions should be drawn from it. These arguments suggest that quantum mechanics is philosophically problematic, and thus that we should defer endlessly to instrumentalism nearly a hundred years after quantum mechanics' inception. Upon investigation the motivation behind this attitude betrays an obvious bias.

Quantum mechanics in fact does not have serious philosophic problems, only *philosophic inconsistencies with the metaphysics of materialism*. Thus when it is argued that "no one understands quantum mechanics," what is really being said is that "no one understands quantum mechanics on materialist grounds." Instrumentalism is then used to avoid confrontation with these inconsistencies such that the issue will never be resolved, leading back in an endless fashion to people once again "not understanding quantum mechanics."

These opinions are extra-scientific and not part of the science itself. As we have seen when the actual science is investigated, it leads irrevocably to an immaterialist picture of the world, and an immaterialist picture of the world which involves the mind, as the observer effect has shown. Here the reader is encouraged to remember that when these sorts of opinions are asserted without evidence they can equally well be ignored without evidence.

B. Quantum Gravity

Perhaps more disturbing than the implications of quantum mechanics are the discoveries arising in the field of quantum gravity. As it turns out matter is not the only illusion. When we examine these discoveries we find out that the very fabric of space-time is as well.

Before beginning, it should be noted that the field of quantum gravity, the quest to merge quantum mechanics with Einsteinian general relativity, is an unfinished program. Several competing theories of quantum gravity exist, the two most significant of these are superstring theory and loop quantum gravity. However, despite the differences between these and various other approaches, it is generally agreed that space-time as we know it is an illusion,[11,12] and that it emerges from information[13] outside of what we would call physical reality.[14]

We will not be looking at all of these theories, as many books have been written on this topic. We will look at some of the basic principles discovered though, and how they lead to the conclusion of emergent space-time. At that point the broader implications of this will be looked at. As we shall see, the physics meshes very well with the Principle of Mentalism.

i. The Holographic Principle: The holographic principle arose from the study of the thermodynamic properties of black holes, and was first mathematically derived in 1998 by Juan Maldacena.[15] Since then the application of this principle to the universe at large has been established on experimental grounds in 2017, through study of the radiation left over from the Big Bang.[16] To understand the holographic principle however it is important to know how it was derived.

The line of research that led to the holographic principle started with the black hole information paradox. The second law of thermodynamics says that entropy always increases over time. In turn this law is seen as one of the most inviolate laws of physics, second only perhaps to the laws of conservation. In the case of black holes it appeared to be violated though. As is commonly known nothing going into a black hole can ever come back out. It would have to exceed the speed of light to do so, and yet relativity tells us that nothing can go faster than the speed of light. Thus any entropy entering a black hole would appear to be destroyed and disappear from existence, seemingly violating the second law.

Then in 1972, Jacob Bekenstein noted that there was a direct correlation between a black hole's size, and the amount of entropy contained within it.[17] Specifically, the black hole's entropy is directly proportional to the surface area of its event horizon, the black hole's surface. This limit is known as the Bekenstein Bound.

There is another limit that is important for comparing to the Bekenstein Bound to arrive at the holographic principle. This is the natural limit to the amount of information that can be extracted from any given region of space set by the conditions of general relativity and the Heisenberg uncertainty principle.

The uncertainty principle tells us that beyond a certain limit the more information we learn about a particle's position, the less determined its momentum will be and vice-versa. By less determined what we mean to say is that it is literally in a superposition of states spread out over the span of values in which it is undetermined. Thus as more precise information is extracted regarding the particle's position, its span of energy values also increases, as energy increases with momentum. Now if the span of energy values increases too greatly within a limited region of space, that region of space will turn into a black hole from which no more information can be extracted.

Thus there is a natural limit to the amount of information that can be extracted from a region of space.

In turn given that entropy is measured in terms of information it turns out that this limit also happens to match the Bekenstein Bound. Meaning the amount of information that can be extracted from a region of space before it turns into a black hole is proportional to the information content of the black hole.

This information content is proportional not to the *volume* of the black hole but to a quarter of its *surface area*. In essence no further information can be stored in the black hole that is not already encoded on its surface. This is quite astonishing as it is parallel to the contents of a closed package being knowable just by looking at the surface of the package. In other words, the contents of a black hole are nothing more than a holographic projection of information stored on its two-dimensional event horizon. Extending this further, we see that this principle applies to regions of space in general and not only to black holes. Any three-dimensional region of space can be described equivalently in terms of the information on its two-dimensional surface and vice-versa. This even applies to the universe at large, leading to the so-called holographic universe, which has been experimentally established in 2017 as mentioned before.

ii. The ER = EPR Correspondence: Though the holographic principle can be applied to study a wide variety of phenomena in a wide variety of contexts, there is one particular solution that has provided a deep clue to the problem of quantum gravity. This solution is popularly known as the ER = EPR correspondence. The EPR here stands for quantum entanglement and is named after Einstein, Podolsky, and Rosen who first posited entanglement. ER in turn stand for Einstein and Rosen who first described wormholes with general relativity.

Here Juan Maldacena and his colleague Leonard Susskind found that if the two-dimensional event horizons of a pair of black holes were entangled, the equivalent three-dimensional space-time described by the holographic principle was that of an Einstein-Rosen bridge, more popularly referred to as a wormhole.[18] For those unfamiliar with the concept, a wormhole is a tunnel through space, connecting two separated regions of space-time via a space-time bridge. In the case of the ER = EPR correspondence, the two mouths of the wormhole are the event horizons of the entangled black holes.

iii. Emergent Space-time: The reason the ER = EPR correspondence is important is that it tells us something fundamental about the nature of space and by extension time, as space and time are linked according to relativity. Space is defined through the relation of separate locations. Without two or more locations there can not be any meaningful definition of space, as every student of geometry knows. However ER = EPR reveals something about the nature of space. As basic geometry tells us, locations define space. Yet ER = EPR shows us that the quantum state in entanglement causes a pair of locations to arise with respect to one another in the form of wormhole mouths. Thus the space between the wormhole mouths emerges from the quantum states in the wave-function, which in terms of the holographic principle are nothing more than quantum information stored on the surfaces of the black holes.

This has led to the realization of a more general principle that space emerges from quantum states in the wave-function. Physicists such as Sean Carroll have noted that space-time emerges from information in Hilbert space,[19] which though not a space in the proper sense, is a domain of reality in which quantum probabilities exist prior to collapse.

The fact that space-time is emergent rather than fundamental leads us to an interesting conclusion about the nature of physical reality. Ever since Kant, physical reality has been defined as those things existing within space and time. Thus space-time's being emergent is the same as physical reality's being emergent. Thus whatever physical reality reduces to it is inherently non-physical. In turn this leads us to an even more astonishing realization.

iv. Digital Physics: If space is not fundamental, then that tells us that space does not actually exist at the deepest level of reality. Furthermore, as seen from this level, what we call space, is actually a system of spaceless quantum information. In other words, space as we think of it as a fundamental thing is an illusion. However, an illusory physical world generated by underlying information is identical in every way with a virtual reality, save for perhaps the physical substrate our virtual realities are based upon. Direct parallels between our world and a virtual reality can even be directly noted in the information processing effects seen in everyday video games. Many of these exist and can be explored by the reader elsewhere. However, for our purposes it is only necessary to list a few here.

1. A Maximum Speed Limit: Any virtual reality will have a maximum speed determined by the processing rate of the computer generating it in any reference frame. Our world has exactly such a speed limit in special relativity with regards to the speed of light.

2. Processing Rates Slowed By Higher Loads: When a virtual reality has to process more information, time will appear to slow down within the game. This has an exact parallel to time slowing down in the vicinity of large masses in general relativity given the equivalence between mass and energy, and that energy can be measured in terms of information content. The objection may be raised that this would slow down the entire virtual reality at once rather than just an area. However, this is explained by the fact that information processing in our world is distributed evenly across space in the wave-function.

3. A Beginning: Every computer game has a start-up point prior to which it was not being simulated. Such an event demonstrates that this virtual reality is not fundamental but derivative from another more fundamental reality. A fundamental reality would not possess such an event. However, our universe did have just such an event at the Big Bang.

4. Objects Render Upon Observation: In video games objects that are not on screen do not exist. Rather they are stored in the hard drive as data until they are rendered to the player on the screen. Likewise, in quantum theory objects do not exist before they are observed or measured.

5. Tunneling Behavior: Due to load processing limitations two objects in the same location can not always be simulated simultaneously. Thus sometimes objects will appear to pass through barriers while the barrier is not being simulated. This is commonly seen in modern video games, and is an inherent feature of them being based on information processing.[20] This phenomenon also shows up in the form of quantum tunneling in our world.

6. Non-Locality: Objects can affect each other instantly in a video game despite being separated by vast distances due to being correlated by the same underlying program simulating the vast underlying distance in the first place. The parallel to quantum entanglement is obvious here and has been discussed before.

This are just a handful of the parallels between virtual realities and the features commonly seen as counterintuitive in modern physics. The computer scientist Brian Whitworth has famously compiled a much longer list of such parallels if the reader wishes to explore this further.[21] The underlying point being made though, is that given the evidence previously examined and the parallels to a virtual reality, it becomes apparent that the world we live in is fundamentally illusory and simulated from information. It should be noted however that it is not being fundamentally simulated by a real or physical reality such as one may infer from philosopher Nick Bostrom's Simulation Argument.[22] This is not to say these sorts of simulations might not exist. However, if they do, then the world simulating our world must itself involve quantum mechanics, and thereby also be simulated.

The reason for this has to do with the difference between quantum and classical information. Classical information stores only a single state per bit, either a 1 or a 0. By comparison, a bit of quantum information, or a qubit, stores many states per bit. So if our world, which possesses qubits, is simulated by a classical world, all of the qubits in our world would need to be represented by individual bits in the "real" world. Any computer that would do this would need eat up both hard drive space and computing power exponentially, making any such computer impossible to build.

So given the ubiquitous present of qubits in our own world, it is safe to say that any hypothetical simulating world would need to have qubits too. And if it has qubits, it is quantum mechanical and thus would have the same sort of non-realism present in our world. Thus it would have to be "virtual" or derived from information also. So we can not account for the physics within a material or physical ontology. In order to account for it, we will need to first examine some basic metaphysical facts.

C. Ontological Implications of Modern Physics

To understand what physical reality emerges from it is first necessary to examine a few metaphysical implications of the physics. To do this we need to first examine the metaphysical implications of the physics in its own right. Then we will need to examine the metaphysical implications of consciousness, and lastly how these implications pertain to the physics.

The first thing we must note is that the physics points to a form of immaterialism. Whatever the world *is* made of, we can know for a fact what it *is not* made of. It is not made of matter or anything else physical, as physicality is contained to within space-time, and yet space-time is emergent.

Secondly, it *is* made of information. This fact tells us something interesting. First though, we should clarify that what is meant by "information." Information is not merely a representation of information, such as the strings of alphabetic symbols comprising words on pages. This is only a representation rather than the information itself, which exists as semantic content known by a mind.

Furthermore, it is not information embedded on any material substrate as one might argue genetic information to be, or information stored on a hard drive. Any material substrate exists within space-time, and yet this information causes space-time itself to arise. Thus this information exists as fundamental and without any underlying substrate, as the physicist John Archibald Wheeler summed up in his explanation of the phrase "It from bit:"

"It from Bit symbolizes the idea that every item of the physical world has at bottom — at a very deep bottom, in most instances — an immaterial source and explanation..."[23]

Information however has a *mental* ontology as it necessarily entails a *subject* that is *informed*. Thus the picture painted from modern physics once the metaphysical nature of information is understood is that mentality lies at the bottom of the physical world rather than the other way around with mind emerging from matter. As we will shortly see, the other metaphysical implications only corroborate this fact.

The last metaphysical implication to be drawn from the physics is the fact that it demonstrates the physical world to be ultimately illusory. If physical reality is an illusion, then it becomes an easy matter to contrast it with what is not an illusion. Whatever is not an illusion must not exist as part of an illusion either and therefore must be more fundamental than it. But here we immediately know that the mind is not an illusion. As Descartes pointed out, the very act of attempting to doubt the existence of a doubter indicates that there is a mind doing the doubting. Thus given that the mind is not an illusion, the mind must be more fundamental than physical reality. When we examine philosophy of mind independently, we find that this conclusion is supported from other facts of consciousness as well.

For instance a simulation itself can not support consciousness. This can be shown from a popular but simple thought experiment known as the Chinese Room.[24] In the Chinese Room, a man is locked in a room with a set of grammar books instructing him how to decipher and respond to Chinese symbols written on pieces of paper that are handed to him from outside the room. However, as he does not know Chinese, his manipulation of the syntax of its grammar can not be said to be identical with the semantic knowledge of Chinese of the fluent Mandarin-speaker on the outside of the room. This is despite the fact that this Mandarin-speaker judges his responses to seemingly represent a fluent understanding of Chinese.

The Chinese Room demonstrates that any sufficiently programmed artificial intelligence or AI can not possess true understanding either. Since the manipulation of syntax as instructed by the grammar books is no different in principle than the manipulation of operations on a computer program, an AI programmed with such a program can not understand Chinese any more than the man in the room. Since any process is at bottom nothing more than the manipulation of some kind of syntax, whether it be cybernetic, grammatical or even neural, this demonstrates additionally that consciousness which possesses understanding can not be a process.

However if this is the case, and by extension consciousness transcends any syntactic program, it follows that it must transcend any virtual reality generated by such a program. Given that we have concluded on grounds of physics that our world is such a virtual reality, consciousness must transcend it. In other words consciousness must come prior to the emergence of space-time, not after it.

In turn when we look at consciousness in relation to the notion of emergence, we see that this fact fits well. If there is one thing we are immediately aware of regarding our own consciousness, it is that it is irreducible. We can not for instance break the sensory perception of blueness down into anything that is not blue, nor if we introspect upon it, can we reduce thoughts into anything other than thoughts. And if something is irreducible it is also fundamental. Reducible things can be broken down into smaller parts, but irreducible things can not.

Some have posited that consciousness is strongly emergent, meaning that it emerges without any explanation from some underlying material substrate. Such an account is by definition magical in the pejorative sense. Something that appears without even a possible explanation or accounting of it in terms of what it is said to appear from exists for no reason at all, and is beyond the boundaries of rationality. Given that the so-called emergence of consciousness can be rejected on grounds of its being necessarily magical, the logical conclusion is that consciousness is fundamental. Though if consciousness is fundamental, yet space-time is not, then consciousness is more fundamental than space-time.

Lastly, this conclusion makes sense of the chief long standing problem in philosophy of mind, the problem of consciousness, alternately called the Hard Problem. The Hard Problem is how it is possible for a purely non-mental material substrate to give rise to mentality, which is described in wholly different terms. Matter is third person and objective, while consciousness is first person and subjective. As they are defined in entirely different terms it does not make sense how one could relate to the other much less account for the other.

If consciousness precedes space-time though, then there is no need to posit a Hard Problem at all. Consciousness does not arise from matter, for matter does not truly exist! Rather the *illusory appearance* of matter arises from underlying consciousness. Given that accounts of consciousness arising from matter are logically excluded from the facts of consciousness described here, it is apparent where the answer lies. We must account for space-time and physical reality in terms of some underlying consciousness.

D. The Physics of the Mental Universe

i. The Physics of Consciousness: To account for the emergence of the physical world from underlying mind we first need an account of consciousness in terms of physics. Candidates for this already exist that would fit the requirement of consciousness existing beneath space-time. These are the Conscious Realism of Donald Hoffman, the Informational Idealism of Daniel Toker, and the more generic approach of quantum cognition. Here we will go over each briefly:

1. Conscious Realism: Conscious Realism attempts to derive physics in a bottom up fashion from simple interacting conscious agents modeled in terms of Markov kernels. This approach has had good success at replicating the space-time of relativity as well as the wave-function of quantum mechanics. Noteworthy to us though is that it shows entanglement to be identical to consciousness.[25]

The math of conscious realism also includes something called the Combination Theorem which demonstrates that systems of interacting conscious agents themselves constitute conscious agents. This will become relevant later.

2. Informational Idealism: Informational idealism is posited by Daniel Toker and comes to the conclusion that information is essentially identical with experience and comes before the material.[26] Here he borrows from the Integrated Information Theory of Giulio Tononi which derives from the simple realization that consciousness is identical to information in the form of thoughts and perceptions, integrated together within a single perspective.

To tie Toker's informational idealism in with the fundamental physics it should be noted that Tononi identifies quantum entanglement as a quantum mechanical form of integrated information.[27] It is also worth noting that when it is combined with relational quantum mechanics, this definition of consciousness as entanglement happens to produce the same result as the Combination Theorem.

Though it is uncertain whether or not this can be translated into a quantum analog. Max Tegmark has argued that quantum integrated information, or quantum phi, can never reach a value greater than 0.25.[28] Though this has been contrasted with Kobi Kremnizer and Andre Ranchin calculating a quantum phi value greater than 1.[29] Until this issue is resolved it is uncertain whether or not this can be integrated with fundamental physics.

3. Quantum Cognition: The field of quantum cognition first arose from the realization that the math behind quantum mechanics better fit the study of economic decision-making. It was then realized this was due to the fuzzy logic behavior of the human mind, which just so happened to behave exactly like states in quantum superposition.[30,31]

Many researchers in quantum cognition will not venture to suggest these similarities are anything other than uncanny parallels, perhaps to avoid the materialist taboos within academia. However, given the facts of consciousness on the one hand, and the problem of simulating quantum bits with classical systems on the other, it is reasonable to take this similarity as literal. In fact in a future chapter we will see how when combined with the facts of emergent space-time, it accounts for another obvious fact about the nature of inner thought, which can be readily verified within the reader's own awareness.

Some may object that these accounts do not fit into the mainstream model of neuroscience, and this may be true. However, it should likewise be noted that the mainstream materialistic models of the mind are categorically incapable of solving the Hard Problem of Consciousness, and thus do not account for the true nature of consciousness.

Secondly, the objections raised regarding the mind not being quantum mechanical due to quantum processes not being able to exist in the warm, wet, and large environment of neural processes has been shown to be severely dated and falsified. In recent years, quantum biology has demonstrated that quantum processes already exist in large, warm, and wet environments within living biological systems. For example, entanglement has been found to guide the migration of birds[32] and photosynthesis has been shown to rely on quantum computing.[33] Quantum phenomena have even been found in the central nervous system such as the phenomenon of quantum smell,[34] and qubits being found within microtubules in the brain.[35]

To date many proposals exist regarding where the quantum mechanism for consciousness exists in the brain. Some such as Hameroff and Penrose have argued that it lies within the microtubules, or within gamma synchrony which has been shown to be orchestrated by quantum entanglement.[36] Still others such as Beck and Eccles have argued that it lies within quantum tunneling processes across dendrite walls.[37] More recently in 2015, another proposal by Matthew Fisher has suggested that the mechanism lies in phosphorus atoms held in superposition in in the brain.[38] This has some evidence based on the study of psychological changes in rats due directly to alterations in their quantum biology.[39]

Given the relative ubiquity of quantum biology, and proposals regarding possible quantum biological mechanisms in neuroscience it is unreasonable to discount quantum mechanical mechanisms for consciousness in the brain. Furthermore, given that materialistic accounts of consciousness can be eliminated prima facie on grounds of failing to meet even the most basic facts of philosophy of mind it is entirely reasonable to suggest that the precise quantum correlate of consciousness will be found in time. Thus this should not pose a serious problem to the idea of consciousness existing as a feature of fundamental physics.

All three of the models given above arrive at the same conclusion: consciousness is identifiable with entanglement, or more generally in the case of quantum cognition, with the quantum coherence which can produce entanglement. Given what we have already pointed out about the nature of space-time being emergent from quantum entanglement, or the quantum state more generally, it follows directly that space-time emerges from some underlying consciousness. That is, our world is an illusory virtual reality being simulated within consciousness, as the physicist Thomas Campbell would say.[40] Here we can see our first direct parallel to Hermetic philosophy:

"The universe is Mental,"

ii. The Combination Theorem: When we explore the combination theorem, or its analog in regards to Toker's Informational Idealism, in relation to the physics of the mental universe we see that in turn it replicates two more Hermetic teachings in regards to The Principle of Mentalism.

According to Conscious Realism when two conscious agents interact in a system, the system itself comprises a third separate conscious agent constituting the both of them, but distinct from either of them.[41] This principle in turn can be generalized to systems of any number of conscious agents.

This same result can be derived from the quantum understanding of Informational Idealism. Classically separate systems of information must be segregated so as to distinguish the thoughts and perceptions of separate individuals. However, if this consciousness is identified with entangled information as described before, and the relational nature of quantum mechanics is brought into play, whether or not these are "separate systems" is relative to one's frame of reference.

From a higher frame, two things can be in superposition together and therefore entangled, but from a lower frame neither sees the other as in superposition. Thus they would represent separate minds on one frame, yet would constitute a single mind in a higher frame. Thus entangled information systems would achieve the same result as with the Combination Theorem.

In either case, any wave-function entangling any system would constitute a single conscious agent. This would naturally include the wave-function of the universe, which has also been shown to exist by recent tests.[42] This being the case the entire universe would be contained in a single mind containing everything, what the religion refers to as God and the Hermeticists as The Mind of The All. Here we find the second half of the Principle of Mentalism as described by *The Kybalion* paralleled by the physics also, where it describes the universe as being:

"-held in The Mind of The All."

This relational nature of conscious agents or conscious states also accounts for another important feature found in *The Kybalion* concerning mentalism. Namely beings can exist within The All without being The All. This is found in warnings concerning "the half-wise" who foolishly identify themselves as God because they happen to be a part of "All." Here "All" is described as being within "THE ALL," demonstrating a distinction between the two, and yet "THE ALL" is described as immanent within All:

"All is in THE ALL. But note also the co-related statement, that: "It is equally true that THE ALL is in ALL."

This makes sense in light of the relational nature of both entanglement and therefore conscious states or agents in these models. "All" would of course refer to all of the separate conscious states and/or conscious agents existing as subsets of everything that exists. "The All" however, refers to the distinct separate conscious agent containing all these interacting conscious agents or entangled quantum phi states from a higher reference frame, namely that of the wave-function of the universe. However from this frame of reference, the content of these lower conscious states would be part of the conscious state of The All, making The All immanent within these lower conscious states as well.

Thus the Hermetic Principle of Mentalism, as well as some of its subtle details, can be shown to be fully replicated by modern physics and quantum models of cognitive science. Given the importance of this principle, the other principles will follow in a simple fashion from this principle. Thus if the reader has followed us thus far, the bulk of the science needed to grasp Hermetics is already understood.

Before we continue, it is worth noting that one does not need advanced physics to derive Mentalism. Here we will briefly show how Mentalism can also be confirmed directly from the subject's own conscious experience in a very simple fashion without the need of any physics. Then we will progress to other principles.

E. Mentalism From Self-Awareness

It is also possible to derive mentalism from the barest facts of consciousness. Once this is explained, the reader will no doubt be astonished at how simple it is to verify the truth of Mentalism from nothing more than the intrinsic awareness of the facts in his or her own consciousness.

The French philosopher Descartes noted the one thing that he knew with absolute certainty was the existence of his own mind. As he noted, this would remain true even if everything else in existence, including anything non-mental, were an illusion. The fact that mind is such that it could exist in the absence of non-mind demonstrates that it is irreducibly mental, and not dependent for its existence in any way on anything material.

Here the materialist will object that the mind is dependent on something material, namely the brain. However, truths known with a priori certainty can never be refuted with a posteriori claims, as a posteriori claims depend upon a priori truths for their validity. Instead a closer examination of this so-called "fact" is called for.

Given that the mind is irreducibly mental, and that the reader can verify this from facts in his or her own consciousness, the question arises as to how it would be possible for anything to act upon it so as to cause an effect on the mind. Interaction in physics is defined in terms of properties the interacting objects share with each other. In turn these properties are a subset of metaphysical properties.

Thus if two things interact they share both physical (that is describable by physics not necessarily existing within space-time), and by extension metaphysical, properties. In turn having metaphysical properties of a certain kind is how substance is defined in metaphysics. Therefore, if mind is acted upon by something else, say the brain, then it follows that given that the mind is an irreducibly mental substance, then so too must the supposedly material brain be comprised of mental substance!

In fact, anything interacting with the mind at all must itself be mental. Thus anything our minds can perceive as part of the world must be composed of mental substance. If it were not mental, it could not interact with our consciousness, and therefore we could never perceive it! Thus the very fact that science relies upon our ability to perceive the world, entails that the entire scientific method rather than being materialistic, relies upon the external world being composed of mental substance!

Thus The Principle of Mentalism can be established on these two facts implicit in the readers' own consciousness: the immateriality of his or her own mind, and the fact that only like substance can interact with like. It follows from these that the only substance we can interact with, including that of the world we naively assume to be material, must be mental. Of course this conclusion is the same as that of the Principle of Mentalism:

"THE ALL is MIND; The Universe is Mental."

III The Principle of Correspondence

After the Principle of Mentalism, the Principle of Correspondence is perhaps the second most important of the Hermetic principles, and helps to establish much of the overarching framework of the Hermetic worldview. This principle is summarized in two simple formulations:

"As above, so below,"

and,

"As within, so without."

The first formulation is found in both *The Kybalion* and *The Emerald Tablet of Thoth*. The second is attributed to Hermes Trismegistus in the following quote:

"As above, so below, as within, so without, as the universe, so the soul…"

The "as above, so below" formulation refers to the phenomena on the physical plane being mirrored in corresponding phenomena in the higher spiritual planes of existence, what are commonly referred to as the "supernatural." Though these levels of reality are certainly beyond what we would refer to as "the physical world," their existence can be established and described with much of the same science as in the previous chapter. This is due to how the principles of Mentalism and Correspondence are closely linked, as will be seen in this chapter.

The "as within, so without" formulation of this principle refers to how manifestations in the exterior physical world correspond to things occurring in the inner world of the mind. This idea may at first seem metaphorical or poetic. However, despite appearances, it has a solid explanation that derives directly from the physics.

The question may also arise as to how these two formulations are linked. The world above refers to the "supernatural" world, whereas the world within, refers only to our inner mental life. These two may seem separate, but are in fact the same. *How* they are the same has a rather surprising explanation: our inner mental realm is *already* beneath space-time. Furthermore, the rest of the realm beneath space-time beyond that of our minds is also an inner mental "space." It is simply that the vast majority of it lies beyond our consciousness in what Jung termed the collective unconscious.

Here we will explore how this principle can be derived from the physics, as well as some of its implications. Additionally, it will be shown how correspondence actually makes sense of a certain fact concerning introspective consciousness. When this is realized, rather than being mysterious, this "supernatural, beyond space-time realm" to which physical events correspond makes immediate sense.

A. The Physics of Correspondence

i. As Above, So Below: In the previous chapter it was explained that space-time is emergent and ultimately an illusion very much like the worlds of computer games. Unlike in today's classical computers though, our universe is the product of quantum computation occurring in the wave-function. In quantum mechanics these wave-functions are described as existing in a mathematical space referred to as Hilbert space. The precise mathematical definition of Hilbert space, is the mathematical space containing all functions who have finite square integrals. The reader may or may not have the background to understand this. However, what must be understood is that this is not a "space" in the classical spatial idea of the word. Rather the word "space" is being understood here to refer to a domain of mathematical functions.

At this point, most physicists will simply treat this as a mathematical abstraction useful for making predictions and nothing more. This sort of thinking is misguided. It is true that the mathematical description referred to as "Hilbert space" is useful for making predictions. What is ignored though is that descriptions always describe realities, meaning it is not *merely* a description.

As was noted in the previous chapter, our space-time emerges from Hilbert space.[1] This is not to say that it emerges from a mathematical abstraction, but rather that it emerges from very real quantum states, which exist in a domain of reality referred to in mathematical terms as "Hilbert space." As space-time emerges from this domain of reality, it can not be described as spatial in the same sense that we speak of things as having spatial extension, direction, dimensionality and so forth. It exists beneath all of these things. Yet it can still be called a domain of reality. The world in which these quantum states exists is very real. Interestingly this sort of description matches closely what *The Kybalion* describes as a "plane" existing beyond the physical world:

"At the beginning we may as well consider the question so often asked by the neophyte, who desires to be informed regarding the meaning of the word "Plane," which term has been very freely used, and very poorly explained, in many recent works upon the subject of occultism, The question is generally about as fellows: "Is a Plane a place having dimensions, or is it merely a condition or state?" We answer: "No, not a place, nor ordinary dimension of space; and yet more than a state or condition. it may be considered as a state or condition, and yet the state or condition is a degree of dimension, in a scale subject to measurement."

Here we see that it is not a "place" in the classic sense of its being spatial. Yet it is also compared to a condition, yet it is not itself a condition. We will see why it is compared to a condition shortly. However, the description given also seems to describe it as being somewhat akin to a "place," even though it is not in the proper sense. This matches very well with the physics notion of Hilbert space existing as a domain of reality, yet not as literal "space" in the proper sense of the word.

It should also be noted that as it exists outside of space-time, or what most people would normally refer to as the "natural realm." This realm is actually synonymous with what is called "the supernatural." Here these terms "natural" and "supernatural" take on a somewhat different meaning than is commonly assumed. They are merely the world within physical reality or space-time and the world beneath it which produces it. Both are explicable in scientific terms, though unlike with the conventional understanding of the term "supernatural."

This should make sense though, if we consider Mentalism. If the world is a mental construction within The Mind of The All, then it follows that there is both a world that is within the mental construct, as well as a deeper layer of reality comprising thoughts in The Mind of The All, which is constructing this world. To borrow from a computer game analogy, there is both the virtual world of the game, as well as the "world" of the program operating on the hard drive of the computer, beneath it.

Here the "world" of the program and the hard drive would be seen as "supernatural" with respect to any characters existing within the "natural" world of the game. Furthermore, the Principle of

Correspondence can be seen in this analogy directly. The "supernatural" code would *correspond to* an object existing within the video game that is being simulated by it. Likewise, anything existing in the game world would correspond back to a program existing in the "world" of the hard drive.

Likewise in the physical world, objects in space-time as well as space-time geometries themselves would *correspond to* patterns of quantum entanglement existing below space-time (or above space-time depending on your preference), in Hilbert space. These patterns of entanglement would of course encode the quantum computations from which physical reality arises. As can be seen, this is simply the statement "As above, so below" described in scientific terms. Corresponding to every natural reality is a deeper supernatural reality.

ii. As Within, So Without: The next step to reconstructing a scientific understanding of Correspondence is to combine the concept of emergent space-time with the concept of quantum cognition. Once this is achieved, it will replicate the rest of the essential framework of the Principle of Correspondence.

"As within, so without" correlates the inner realm of the mind with the realm behind physical reality. Of course the realm of physical reality is that of space-time. Behind it, is what space-time emerges from, namely the realm of Hilbert space. Hilbert space is of course quantum probability space, the domain of reality in which quantum states exist in superposition.

Quantum cognition meanwhile shows that our inner world of thoughts and emotions has a precise mathematical correspondence to quantum states in superposition. This is seen in the fuzzy logic in decision making. When we purchase an item for instance, we often vacillate for a while between several choices, even after all of the relevant information about them is known.

This sort of vacillation is not what we would expect if our minds behaved as classical computers operating on simple binary logic. It does resemble the processing of states in quantum superposition though. When two things are in superposition together in a quantum state, they are not determined in a deterministic fashion as with a binary computer, but in an irreducibly probabilistic fashion described by the Schrodinger equation. These probabilities begin to weight in either direction, but are never actually decided until the collapse of the wave-function.

This is more than a vague resemblance. As cognitive scientists have discovered, our inner cognitive processes model as though they are quantum information processing in the wave-function.[2] This means that quantum cognition describes thoughts and emotions in terms of objects existing in Hilbert space. Of course here the term "Hilbert space" could simply refer to a mathematical modeling, and the parallels with quantum mechanics could be purely coincidental.

Given what was discussed in the previous chapter on Mentalism though, this seems not to be the case. Furthermore, as explained earlier, if any system that behaves as though it were quantum mechanical were broken down into a classical system, the classical system doing the processing would slow down exponentially with each decomposed superposition. This combined with the basic a priori facts of consciousness would indicate that this is more than a modeling. The internal states of the mind would be quantum states in Hilbert space.

The implications of this are profound. If our introspective mental contents exist as quantum states in Hilbert space, then the world beneath space and time is no mystery to us at all! In fact, you the reader have direct access to this world even as you speak. It is literally as close to you as your own thoughts and emotions! Thus your inner mental life becomes part of the world beneath space-time!

It is at this point that the phrase "As within, so without" begins to make sense. The exterior "physical" world of perception is simulated by the inner world of thought and emotion. Moreover, the world "within" is the same as the world "above." Man's soul is part of what has been traditionally called "the supernatural." This of course makes sense of the statement attributed to Hermes:

"as the universe, so the soul..."

In essence outer space emerges from inner space. This neatly parallels the belief of the ancient Egyptians that man is a reflection of the cosmos. It also on the grand scale explains the esoteric but literal truth behind the common religious concept of man being made in the "Image of God." Man's soul is a reflection of the cosmos, which in turn given the Principle of Mentalism is a projection of The Mind of The All. Man's soul is in essence a miniature reflection of God's mind.

Of course at this point two questions arise. Firstly, why are we not aware of everything in the exterior world if our own inner consciousness is what produces it. Secondly, why are we not able to influence it with our minds?

The first of these questions will be explored later in this chapter. The answer has to do with what the psychologist Carl Jung termed the "collective unconscious." The answer to the second is that we do in fact influence the outside world, but this effect is very tiny and goes mostly unnoticed by the individual. This influence is of course related to so-called psychic phenomena, which despite the protestations of materialists and pseudoskeptics has solid experimental proof with many studies to support it. But this will be explored later in the chapter on the Principle of Cause and Effect. For now though, we should return to our study of Correspondence.

Earlier it was noted that the term "plane" used in *The Kybalion* was comparable to a "state or condition" in addition to a dimension, but that it was neither in the strict sense. The reason for this becomes apparent. Our thoughts and emotions are often described as "conditions of the mind" or "mental states." Here our thoughts and emotions in addition to being our own, are also part of the rubric behind the physical universe, much as a slice of geologic strata reveals the layers beneath a sliver of ground. Thus this domain of reality could also be conceived of as a "mental condition."

Of course someone may argue that just because our thoughts and emotions are quantum states, does not mean that all quantum states are thoughts and emotions. Nor would they argue that this means that the domain of reality in which quantum states exist is necessarily a mental or spiritual reality.

While it is true that thoughts and emotions happening to be quantum states does not necessarily entail the converse, it is also the most reasonable view, and it can be reached from another line of logic. Firstly, it should be noted, that nothing makes the quantum states constituting thoughts and emotions different in any fundamental way from other quantum states in superposition. Thus it would be ad hoc to create such a distinction. Secondly, if that is what thoughts and emotions are in their physical nature, and quantum states are not dissimilar to each other in their fundamental nature, then logically it should follow that these other quantum states are also mental states, as would be implied by Mentalism.

iii. Three Great Planes: It is at this point we can account for the Three Great Planes. *The Kybalion* references these planes as follows:

"For the purpose of convenience of thought and study, the Hermetic Philosophy considers that the Universe may be divided into three great classes of phenomena, known as the Three Great Planes, namely:

I. The Great Physical Plane.
II. The Great Mental Plane.
III. The Great Spiritual Plane."

Starting from an understanding of correspondence we can see what these would refer to. The Great Physical Plane of course references the virtual reality we call "physical" and the field of perceptions collapsed from the wave-functions within it.

The Hilbert space domain beneath space-time would then be split into the contents of the mind; thought and emotion. The Great Mental Plane refers to the subset of these contents constituting thought, while the Great Spiritual Plane refers to the subset of these contents constituting emotion.

The extension of our minds into these three planes was understood by the ancient Egyptians. These three planes and more specifically the subject's stability within them was symbolized by the Djed pillar, the representation of the "spine of Osiris." The Djed was a pillar with three consecutive caps one on top of the other, creating three divisions, one for each plane. This draws a clear parallel between the Hermetic teachings in *The Kybalion* and those of ancient Egypt.

iv. Vibrational Planes: The Three Great Planes are not the only way the term "plane" is used in *The Kybalion*. Another description is given as well that has relevant parallels to physics:

"A "dimension," you know, is "a measure in a straight line, relating to measure," etc. The ordinary dimensions of space are length, breadth, and height, or perhaps length, breadth, height, thickness or circumference. But there is another dimension of "created things," or "measure in a straight line," known to occultists, and to scientists as well, although the latter have not as yet applied the term "dimension" to it–and this new dimension, which, by the way, is the much speculated about "Fourth Dimension," is the standard used in determining the degrees or "planes."

This Fourth Dimension may be called "the Dimension of Vibration." It is a fact well known to modern science, as well as to the Hermetists who have embodied the truth in their "Third Hermetic Principle," that "everything is in motion; everything vibrates; nothing is at rest." From the highest manifestation, to the lowest, everything and all things Vibrate. Not only do they vibrate at different rates of motion, but as in different directions and in a different manner. The degrees of the "rate" of vibrations constitute the degrees of measurement on the Scale of Vibrations–in other words the degrees of the Fourth Dimension. And these degrees form what occultists call "Planes." The higher the degree of rate of vibration, the higher the plane, and the higher the manifestation of Life occupying that plane."

Here we must interject something concerning *The Kybalion's* use of the term "fourth dimension," as it is not actually accurate in regards to physicists' understanding of the term. The fourth dimension of physics refers either to time, or to a fourth spatial dimension found in Kaluza-Klein theory and more generally in superstring theory. Though to its credit, *The Kybalion* mentions that this is not a dimension in the spatial sense of the word.

The reasons for this are historical. At the time of its writing, Einstein's Theory of Special Relativity with its use of a fourth dimension had just come out, and was popular with the public. Moreover, prior to this, the notion of a fourth dimension had become popular in the latter half of the 19th century with the work of the mathematician Bernhard Riemann. Due to its popularity, the concept was popular in occult circles as an explanation for psychic and paranormal phenomena. Despite the inaccurate usage of language common at that time, the underlying concept it conveys is valid and has noticeable parallels in modern physics.

In quantum mechanics, quantum states are expressed as wave-functions, which exist in Hilbert space. In turn space-time emerges out of the quantum states existing in this Hilbert space, and does not exist fundamentally. These wave-functions also have their own vibratory frequencies. However there is no limit to the frequencies that could in principle exist in Hilbert space.

This raises the possibility that there exist other domains of vibration in this Hilbert space that could in principle encode entirely different space-times other than our own. Though this is exactly what *The Kybalion* tells us. Other planes exist at higher or lower vibrational domains than our own universe.

In fact, just such a possibility has been recently proposed in physics to account for the phenomena of dark matter and dark energy. Dark matter is of course the hypothesized invisible matter that produces seemingly sourceless gravitational effects on astronomical bodies.

The theoretical physicist Erik Verlinde has proposed a concept known as Modified Newtonian Dynamics, or MOND. According to MOND, the gravitational influence associated with dark matter actually derives from the vibrations of underlying degrees of freedom coming from beneath emergent space-time.[3] These degrees of freedom are how entropy is defined in physics, and modern theories of quantum gravity describe gravity as a result of quantum entropy. In turn MOND has accurately predicted the ratio of matter to dark matter, providing good predictive support for the theory.

Of course this usage of the term "degrees of freedom" is only a very abstract way to describe what is going on beneath space-time. It would be valid to alternately describe them as "differences between things" which have their own associated vibrations. In other words, they are things existing in other vibrational planes, just as *The Kybalion* states. What is called "dark matter" then is actually only the effect of these other planes leaking over into our reality by producing gravitational influences in our space-time.

As *The Kybalion* notes, these planes are populated with all manner of beings considered to be "supernatural;" angels, archangels, demons, even the so-called "gods" of ancient legend, as well as many more supernatural beings reputed to exist in religion and myth. In turn the existence of these planes could easily account for accounts of heaven or hell and other such otherworldly realms.

Here it should be cautioned that the mere existence of these beings does not mean that they should be worshipped. Even "gods" are finite beings with only a finite part of goodness. In comparison to The Mind of The All both "gods" and men alike are still like little children. Only God, The Mind of The All, contains all good. Thus only God should be worshipped, not any dependent being.

Likewise, one should be wary of any being claiming to be a spiritual guide. Many beings exist, including evil beings, called variously the djinn, the demons, or the archons, among other names. These beings would gladly seek to misguide human beings under the guise of being guides and teachers. Only God should be sought to guide us.

B. Jung's Collective Subconscious

No account of Correspondence given to a contemporary audience would be complete without a mention of the depth psychology of Carl Jung. Jung's depth psychology resolves an important problem associated with the Problem of Correspondence mentioned earlier. It also provides a relatively modern way to understand the concepts of correspondence.

Jung's depth psychology can largely be summarized in terms of three essential concepts:

1. The Collective Unconscious: The collective unconscious is a domain of the subconscious common to everyone. The idea here is that in addition to personal subconscious influences, we also have a shared subconscious mind, even though we have separate conscious minds. Thus the same subconscious influences can appear in multiple people despite their being seemingly disconnected and coming from disparate social and cultural backgrounds.

2. Archetypes: Archetypes are those images and patterns which exist in the collective unconscious, which sometimes manifest separately in seemingly independent people's minds. These can come in various ways including through fiction and mythology. Though in Jung's work in dream analysis he primarily identified them in dreams.

3. Synchronicity: Synchronicities are coincidences that defy probability yet have personal meaning to those who have them. Often they are directly related to or involve archetypal material manifesting in the collective unconscious appearing in the external world. The topic of synchronicity relates the Principle of Correspondence directly to the Principle of Cause and Effect. Therefore we will explore it in more depth in the chapter on Cause and Effect.

The collective unconscious is our first examination of a phenomenon sometimes referred to as "paranormal" or "occult." Though of course it has a perfectly rational explanation and is not at all "magical." It only appears that way to a naïve view of the world.

Before the question was raised as to why we do not have omniscient knowledge of the physical world if the physical world emerges from the inner mental world. The answer is that while our minds are extended in the inner mental world, they do not encompass it. Most of it lies in our subconscious.

This is not merely our personal subconscious either. Rather it is an interpersonal subconscious, what Carl Jung termed the collective unconscious.[4] Since we do not have direct access to the collective subconscious we are not aware of all of the information that is generating the world. Nevertheless, the contents of this collective unconscious can influence our conscious minds unawares.

A very astonishing yet perhaps politically incorrect and crass example of this effect was recorded by Jung from one of his psychiatric patients. This is the famous case of the "Solar Phallus Man."[5] One of his patients, had a strange dream with archetypal imagery of God's phallus hanging from the sun, bringing life to the earth by ejaculating on it.

This story circulated among Jung's colleagues who found this to be humorous. A short time later, the contents of an ancient Mithraic text were published for the first time describing identical archetypal symbolism as in the patient's dream. Furthermore, Jung was quite certain that the patient had never seen this imagery or known of this text before. The only way to account for the contents of the patient's dream was through archetypal material bleeding through into his dream from the collective unconscious.

Such an account appears bizarre and beyond scientific explanation, but Jung had recorded many instances like this. Before it is dismissed as outside of science, it should be noted that the physicist Wolfgang Pauli, who discovered the neutrino, believed it could be accounted for with quantum mechanics. Pauli and Jung maintained a twenty-six year personal correspondence, discussing these concepts and how they could be accounted for with the then emerging discoveries in quantum theory.[6] If the reader wishes, he or she may examine the book *Atom and Archetype,* which provides a complete record of their correspondences. Unfortunately, at the time, there were not yet sufficient scientific advances for them to complete their project. However, the state of modern physics is significantly more developed than it was in their era.

If we model Hilbert space with quantum cognition it is possible to provide a fully scientific understanding of this collective unconscious in light of the physics behind emergent space-time and by extension Mentalism. Our minds can be described as quantum states in superposition behind space-time. Though they are separate, the rest of Hilbert space is comprised of mental states ultimately encoded in the universal wave-function as well. These mental states would comprise the contents of the collective unconscious. The entire universe then is generated as an informational construct from this collective unconscious.

Separate people's minds would temporarily entangle with these quantum states stored in the collective unconscious. This would be enough for them to become aware of the archetypal material bleeding up from the collective unconscious. Due to the non-local nature of this entanglement, and of the non-spatial nature of the realm beyond space-time, distance would not be a problem. People as distant as the United States and Korea could pick up the same themes in their dreams independently, despite never having seen them in the waking world. Intriguingly, in the field of quantum cognition, Jungian archetypes are sometimes modeled as superpositions in qubits, corroborating the general hypothesis.[7]

As an aside, it is interesting to note that not all of this archetypal material necessarily comes in the form of imagery either. Just as our own minds can be described in terms of quantum information, some of this information can also encode minds!

Of note here is Jung's description of his encounters with a character he named "Philemon." Philemon would appear in his dreams as an old man with kingfisher wings and appeared to have a distinct personality and a mind of his own. As Jung comments:

"Philemon and other figures of my fantasies brought home to me the crucial insight that there are things in the psyche which I do not produce, but which produce themselves and have their own life. Philemon represented a force which was not myself. In my fantasies I held conversations with him, and he said things which I had not consciously thought. For I observed clearly that it was he who spoke, not I. He said I treated thoughts as if I generated them myself, but in his view thoughts were like animals in the forest, or people in a room, or birds in the air, and added, "If you should see people in a room, you would not think that you had made those people, or that you were responsible for them." It was he who taught me psychic objectivity, the reality of the psyche. Through him the distinction was clarified between myself and the object of my thought. He confronted me in an objective manner, and I understood that there is something in me which can say things that I do not know and do not intend, things which may even be directed against me."[8]

The collective unconscious has many such beings. The point of bringing this to the attention of the reader is not so that you should seek out these sorts of encounters, but merely to raise awareness of them. Dreams sometimes contain warnings or advice sent from messengers. Other times they may be produced by negative forces seeking to condition the subject with fear. In fact, the seemingly mysterious phenomenon of "night terrors," difficult to explain through materialist science, is an example of this.

The human soul is caught in the middle of a great battle between these forces, which he or she would be completely oblivious to if operating from a materialistic understanding of the world. Thus the reader should not be ignorant of these facts.

C. The Hermetic Kabbalah

At this point the reader should have grasped enough concepts to understand the general concept of the Hermetic Kabbalah, the so-called grand unified theory of occultism. As the name suggests, this is a combination of Hermeticism and Kabbalah.

Kabbalah of course is the Jewish esotericism, while Hermeticism as explained before derives from Egypt. These two schools of thought form the two central pillars of Western esotericism. Rather than being an ad hoc combination, they are actually quite compatible in much the way a hand fits into a glove.

The best analogy for the relationship between the two is that of software and hardware. Hermeticism is primarily defined by the metaphysics of its two main principles, Mentalism and Correspondence, which we have already gone over here. This forms the "hardware" as it were, the essential metaphysics, which as has been shown can be converted entirely into physics. And the analogy to hardware is not far off either. Hermeticism's contribution is that it explains how the "virtual reality" of the mental universe operates, and "where" the "hard drive" is located in the inner mental world.

Kabbalah by contrast is the "software" aspect of the Hermetic Kabbalah. As any student of the Kabbalah is familiar, the Kabbalah is centered around the concept of the Tree of Life. The Tree of Life is an emanation map which charts emanations in the form of ten "sephirot," existing in a hierarchical chain of emanation through four levels of reality. These sephirot begin with the Three Supernal Sephirot in the realm of God and proceed down through seven lower sephirot with the lowest, Malkuth, being the physical world. In turn, these sephirot are connected by a network of twenty-two pathways describing their interrelations, with each associated with one of the twenty-two letters of the Hebrew alphabet.

It is interesting to note that the Kabbalists view the world as being both spoken into being and continuously upheld by God's continuously speaking words "written" with these letters. Two parallels automatically spring to mind, one with the concept of a virtual world, and the other with direct religious significance.

Firstly, the parallel of words both creating and continuously generating the world has the obviously direct appearance of a computer program running to produce a virtual world. Secondly, throughout the Abrahamic traditions there is a strong notion of God creating the world through speech. This corroborates that doctrine first in terms of the occult framework, and then on scientific basis grounding that framework.

It would not be wrong to think of this Tree of Life as a map of the "program of archetypes" generating the virtual reality. In fact one Kabbalist text, the *Sefer Yetzirah* or "Book of Creation," appears to provide a very good description of a twenty-two level Bloch Sphere:[9]

"These twenty-two letters, the foundations, He arranged as on a sphere, with two hundred and thirty-one modes of entrance. If the sphere be rotated forward, good is implied, if in a retrograde manner evil is intended. For He indeed showed the mode of combination of the letters, each with each, Aleph with all, and all with Aleph. Thus in combining all together in pairs are produced these two hundred and thirty-one gates of knowledge."

The Bloch Sphere is a mathematical representation of the state space of a qubit. Here states are arranged on the surface of a sphere in pairs with an axis for each pair of states. Pure states, meaning single states, are on the surface of the sphere, whereas mixed states or permutated combinations of the pure states are located within the sphere.[10,11] Each such permutation corresponds to a different superposition of states, which would process a different probable outcome within the quantum state.

This is again what is seen in the description in the *Sefer Yetzirah*. The twenty-two letters are arranged on the sphere, producing two hundred and thirty-one permutations corresponding to entrances to within the sphere, the mixed states in the interior of the Bloch sphere. In turn the description of each of these two-hundred and thirty-one different "gates of knowledge" corresponds nicely to the different outcomes in information one would expect from different superpositions in a qubit.

Likewise operations are done on the qubit through rotations of the Bloch sphere, with different rotations corresponding to different outcomes.[12] This appears to correlate with the example given of evil intent and good intent alternately corresponding to the different rotations of the sphere in the *Sefer Yetzirah*.

Though only one example, these peculiar similarities to quantum computing would seem to describe "letters" in superposition in Hilbert space. Given that the context of this passage is in describing the creation of the universe with these letters, this would seem to fit firstly with the general science described before behind Hermetic Correspondence, as well as with the synthesis of Hermeticism and Kabbalah described here.

D. Correspondence from Platonic Intuition

At the end of the previous chapter we saw how Mentalism could be derived from the basic facts of consciousness such that it could be recognized as true on intuitive grounds rather than through complex and advanced physics. Likewise it is possible to grasp Correspondence in a similarly intuitive fashion by expanding upon the intuitions discussed in the previous chapter.

We begin with the facts of consciousness leading readily to Mentalism. The immateriality of the mind derives directly from the fact that the mind's existence is known with certainty, but the supposedly non-mental world is not. This establishes that the mind can not be a part of the non-mental world, and thus that it is necessarily immaterial. However, just as it is not logical for mind to come from non-mind, so too was it shown that it is not logical for mind to interact with non-mind. From this it followed that the world we interact with must ultimately be mental as well, corroborating the statement that "the universe is mental."

One need only to add the intuition of Platonic forms to this, and their relation to Mentalism to readily intuit Correspondence. Platonic items are immediately intuitive, as they are readily recognizable as the ideas we have in our own minds.

Now if we think about them we notice something curious. Ideas are not located in space or time. It makes no sense to say that the natural numbers or the concept of love are to be found in a particular spatial location, nor are they to be found at a particular point in time and no other. Even in our brains we see nothing but neurons and their activity, not the ideas themselves.

The fact that they are not in space and time is a very interesting clue. We have already run across things existing but "not in space or time" before. In fact all of the physics behind both Mentalism and Correspondence is based on emergent space-time, the idea that space-time is not fundamental and that there are things that exist outside of it from which it emerges.

When we examine this, we see that these two sorts of existences outside of space and time are in fact one and the same. Starting from Mentalism we see that the world is a construction within God's mind, but of what is this construction made of? It is of course God's thoughts, the Platonic forms! But "where" is this mental construct being generated from? Outside of space-time in Hilbert space of course!

Thus the "outside of space-time" of the Platonic forms is the same as the "outside of space-time" associated with the emergence of space-time. The only apparent difference between the two is that the latter is often associated with a lack of understanding of the metaphysical implications of the science.

Furthermore, it should be noted that the Platonic forms are singular. There is for instance only one "Platonic form of the number four." Thus when we possess Platonic forms in our own mind, they are the same Platonic forms possessed by The Mind of The All. Of course The Mind of The All goes far beyond ours, but nevertheless our thoughts overlap with those of The Mind of The All insofar as we have them. Thus our thoughts exist in the beyond space-time realm generating our world as well.

From this The Principle of Correspondence can easily be intuited from the very nature of Platonic forms as archetypes. When we speak of a Platonic form, we are speaking of a Platonic form of something. The Platonic form of blue for instance exists both outside of space and time and in our inner mental state. In the external physical world it finds itself manifested in the blueness of the sky, the color of the ocean, or in the pigment of a bluejay or hyacinth macaw. When we recognize this we say that the blueness of the sky "corresponds to" the "Platonic Form of Blueness."

When we recognize that the physical world is a mental construction produced by the Platonic realm, we see that this is more than merely a correspondence. The forms in the Platonic realm in fact construct what is seen in the physical world. This concept is known as emanation and is central to the philosophy of Neoplatonism, perhaps a third pillar of Western esoteric thought. Here, the blueness of the ocean is an *emanation* of the Platonic Form of Blueness.

Once the mental nature of the world is properly understood, these correspondences can also be seen to go beyond the mere one to one correspondences of Platonic forms to their physical instances. As noted by Jung, the archetypes in the inner world are related to each other in more complex ways as well, that may at first appear to be acausal. Ravens are associated with death for instance, or four-leafed clovers with good luck.

These associations are mental in nature, and if as is sometimes the case, these sorts of associations are grounded not in the human mind but in deeper archetypes in the collective unconscious, then these associations sometimes manifest as seeming "coincidences" in the world. This fact is exploited in what is often termed the "interpretation of signs" or "divination." However, this is one of many interesting phenomena associated with Mentalism and Correspondence, which will be explored in the next chapter.

IV The Phenomena of Mentalism and Correspondence

Before continuing on into the Principle of Vibration, it is worth taking the time to look at some of the phenomena associated with the Principles of Mentalism and Correspondence. After all, it is pointless to focus only on theory if that theory is never used to account for actual phenomena. Furthermore, some of these phenomena are quite interesting and will expand the reader's grasp of both the level of significance of these principles, as well as deepen his or her understanding of the larger world.

Given the occult or paranormal nature of many of these phenomena, they are often dismissed by academics. As the reader will discover in a least a few of these cases, this is not for good reason. Many of these phenomena have been well documented under laboratory conditions and established beyond the boundaries required of scientific proof. Others such as out of body and near death experiences have published studies conducted on them. The comprehensive surveys of accounts of these reports indicate firstly that they are not merely anecdotes, and secondly that they have veridical features which exclude them from the possibility of materialistic explanation.

It should be noted here that the fact that they can not be explained in materialistic terms does not mean they can not be explained in scientific terms. This presupposes a completely unwarranted and groundless conflation between science and materialism that rests only on a naïve intuition about the world and not on reason or evidence. In each case, we will attempt to demonstrate how they can very easily be accounted for in terms of the mentalistic science previously described.

A. The Astral Plane

This collective inner mental plane behind space-time described in the last chapter appears to be mysterious and strange, but in reality it is not. Likewise, undoubtedly most of the physicists who use Hilbert space merely as an abstract description of non-local physics do not recognize its true significance, and would probably reject it if they knew what it was really describing. This realm is none other than the so-called "astral plane."

At this point a note should be made on nature of what is often termed "astral projection," or the "out-of-body-experience" or OBE. These two terms are often used synonymously but are in fact different. Astral projections are only a subset of OBEs. What is commonly thought of as an astral projection where the individual is able to move around in the physical realm and see his or her body in real time is not an astral projection but an *etheric projection*. The nature of the astral projection is a bit different and involves the subject being able to leave the physical plane entirely.

It is interesting to note that some physicists exploring the cross section of digital physics and consciousness have already come to a conclusion regarding the explanation for such phenomena similar to that of the authors. Thomas Campbell, a physicist who has worked both for NASA and the DoD, has a similar view of these phenomena as the authors. He arrives at the same conclusions in regards to the Principle of Mentalism, viewing the universe as a virtual reality being contained within what he refers to as the Larger Consciousness System, what the Hermeticists refer to as The Mind of The All.

Beyond the realm of the mental construction we erroneously call the "physical universe," is what he and fellow digital physicist Ed Fredkin refer to as *Fredkin's Other*. This is the realm from which the universe is simulated and is identical with the Hilbert space explored in the last chapter. Along with Robert Monroe, Tom Campbell and others at Monroe Laboratories explored the astral plane. Here

Campbell claims to have in fact visited many other realms within Fredkin's Other. The interested reader can find these accounts in more detail in Campbell's book *My Big TOE*. These are of course the same as the vibrational planes described in the previous chapter.

The explanation for this is actually quite simple and not "paranormal" at all once properly understood. Just as the physical world is a virtual reality, the physical body is an avatar for the mind. Thus the body does not produce consciousness, rather it houses it as a vehicle or interface for the physical world. In some ways this should be obvious. After all, we can not know of the brain's existence from immediate awareness, but we can know about the existence of consciousness from immediate awareness. The two are not the same. You are not your brain, but you do possess a brain.

This being the case, so called "astral projections" can be easily accounted for by the mind leaving its physical avatar and exploring the realm behind the virtual reality from which it comes. Given the description of the collective inner mental plane before some of the details described in these experiences should come as no surprise. People in astral experiences have reported being able to move great distances instantly by merely thinking of another location. Likewise, they also report that their normal spatial and temporal categories do not apply in the astral realm. This would of course be expected of a realm "beneath space-time."

These phenomena, though they may correlate to abnormal brain function, are not accounted for it either, as is sometimes popularly misconceived. Firstly, it should be noted that in the case of the most common kind of out-of-body experience, the near-death experience, or NDE, the brain in fact has no significant activity of any kind. Furthermore, as will be seen in the next section on near-death-experiences, many of these experiences have a veridical quality. Specifically, many of these accounts include details reported by the subject that he or she could not have otherwise known, and that were at the same time corroborated by independent third parties. These are not isolated anecdotes either, but in many cases are the tabulated results of many cases reported in comprehensive studies.

The astute student of neuroscience will raise a very important objection at this point. When the brain decays the mind also decays. Thus a causal relation exists between them. The argument then is that as brain function gradually decays, the mind eventually decays to nothing, until upon death there should be no consciousness at all. Likewise assuming this causal relation, minds should not be able to operate independently of bodies. Thus the more general case of the OBE is also argued to be impossible.

While the observations underlying this objection are accurate, they miss an important but hidden detail. This detail involves a whole new level of structure, and appears as a direct consequence of mentalism without which it would otherwise be overlooked. Once the nature of this structure is understood, the resulting picture accounts for both the causal relation of mental phenomena to neural activity, as well as the veridical nature of many OBEs and NDEs. This will be explored in the next section.

B. Near-Death Experiences

More widely known than the out of body experience is the near-death experience, or NDE. Though the latter is a subset of the former, the NDE is more widely known and is often considered in its own right. Here we wish to explore the evidence for veridical NDEs, and why they can not be accounted for on materialistic accounts. However, it will be shown how they can readily be accounted for with Mentalism. Once this is understood, it will also show how these accounts can be reconciled with the observed dependency of mind states on brain states. When the exact nature of this is understood it will bring two other things to light. Firstly, it will become apparent that this structure was widely known about in the ancient world and has simply faded into obscurity over the subsequent millennia. Secondly, it will reconcile seemingly incompatible evidence concerning the afterlife.

i. Veridical NDEs: The claim that near-death experiences are only an illusion resulting from unusual brain activity is common. It is also easy to refute. Much of the time the accounts given are rejected as merely anecdotal. Upon examining the literature on the topic it becomes quickly apparent that this is not the case. It is important to illustrate with a few examples here:

1. Blind People Experience Sight in NDEs: In a study conducted on 31 cases where blind individuals reported near-death experiences it was found that they all reported the ability to see.[1] Half of them were blind from birth and had no prior concept of vision. What is remarkable is that vision like the other senses requires the brain to organize neural connections in order to process it. People can not simply go from blind to full sight without developing the neural architecture to process such connections. Yet despite this, the people surveyed in this study were able to experience vision with no correlating neural structure to host it.

This ability to gain new capabilities after the brain has shut down despite the apparent causal relation between brain and mind can be accounted for within a mentalistic account of these experiences. This will be explored shortly in the next subsection.

2. Corroborated Information in NDE Reports: In many cases subjects will report back detailed knowledge from outside of the room. In these cases, the doctors knew the patient was comatose or had been declared dead. In one instance, a nurse had removed a man's dentures while he was comatose and placed them into a crash cart. A week later after coming out of his coma, he was able to identify the location of his dentures and that the nurse had taken them. He also gave an accurate detailed description of the room where they had attempted to resuscitate him.[2]

Far from being rare, cases of this sort are actually quite common and can be found in the literature on near-death experiences. Janice Holden of the University of North Texas has tabulated 107 reports of this nature all found in published literature.[3] Given that these accounts are both documented in published literature and are comprehensively examined, they can not be considered as anecdotal.

3. Increased Mental Function Without Corresponding Brain Activity: Lastly it should be noted that accounts of this nature happen while brain activity is either lacking or at the least greatly impaired. By contrast on a materialistic understanding, mental activity should be greatly diminished or non-existent altogether. The famous case of Eben Alexander comes to mind.[4] Dr. Alexander, a prominent neurosurgeon with a materialistic outlook, had extremely vivid experiences at a time when he was comatose and his brain was almost completely shut down. Despite his scientific training specific to this field he ruled out his experiences as being explicable in materialist terms.

Again this marked anti-correlation between increased mental activity and decreased brain activity is not what one would expect from a materialistic framework. Something not being explicable in materialist terms does not mean it can not be explained scientifically however. In fact, these can easily be accounted for in the context of Hermetic and occult concepts synthesized within a scientific framework. This is what we will look at next.

Anomalies of this nature are common in reports of NDEs and defy materialistic explanation. What we have written here is only a small sampling of these anomalies. The reader can find a comprehensive and well referenced account of this subject in J. Steve Miller's *Near-Death Experiences*.

ii. The Binary Soul Doctrine: All of these experiences in the previous two sections are explicable within a Hermetic view of the world using concepts associated with the Principles of Mentalism and Correspondence. These experiences do not fit into a materialist framework, and they are seen as existing but as supernatural and inexplicable within a dualistic framework. The explanation of these phenomena can be found in what is sometimes termed the *"binary soul doctrine.*[5]*"*

Though this has become obscure in modern times, the ancients had a concept of the soul existing as a two-leveled structure; spirit and soul. This can even be found in the Bible today. The earliest reference to spirit and soul can be found in Genesis in the creation account of man:

"And the Lord God formed man of the slime of the earth: and breathed into his face the breath of life, and man became a living soul." –Genesis 2:7

The word used for breath, *pneuma,* is a direct reference to the spirit. This verse describes a direct relation between the spirit, the soul, and the body. A part of the spirit enters the body and a soul is produced. A study of the words "spirit" and "soul" will reveal this same pattern throughout the Bible. This concept is not contained to one religion either, but was actually widespread throughout the ancient world. The Egyptians referred to the spirit and soul as the *ba* and the *ka*, the ancient Greeks referred to them as *psuche* and *thumos*, and the Hindus referred to them as *asu* and *manas*.[6] All of these are simply terms for the two halves of the binary-soul respectively; the spirit and the soul. The similarity in these doctrines seems to indicate an ancient *prisca theologia,* largely obscured today, from which they are derived. The astute observer may notice that these two concepts parallel our notions of conscious and subconscious minds today.

Considering the Principle of Mentalism, all of this makes natural sense. The mind in the form of the spirit predates the body and would exist behind the virtual reality in Hilbert space as per the science of quantum cognition described earlier. A small portion of it is projected into the body within the virtual reality we mistakenly refer to as "physical." This body of course is no different than a digital avatar in a computer game. When the digital avatar is "killed," the player by no means dies. The portion of the spirit projected into the avatar is then what we call the soul.

This concept makes no sense in a dualistic view where there is a material world and a mental world coexisting with no need for an additional layer to the mind. Nor does it make sense in a materialist picture, where, as Hermes prophesied about the future materialistic age:[7]

"As for the soul, and the belief that it is immortal by nature, or may hope to attain to immortality, as I have taught you; all this they will mock, and even persuade themselves that it is false."

It does however make natural sense in the mentalistic picture. If the material world is but a mental construct within The Mind of The All, then it would make sense that there would be a layer of mind within it, and a layer of mind outside of it. The layer of mind within it would exist within and operate through the virtual avatar of the body.

When put into the context of the physics describing mentalism and the world existing behind the virtual reality, this makes immediate sense of the mind's correlation to the brain. That part of the mind which exists in the brain is the soul, and it would be comprised of many entangled quantum states. As the brain breaks down, these entangled states would break down as well. Thus mental capacity would diminish as the brain decays.

Once these states leave the virtual reality though, they could re-entangle with the spirit at a level subconscious to us now, not inhabiting the avatar outside of space-time. Once this happens, given that entanglement can be seen as identical with consciousness, all of the information and mental faculties lost upon the decay of the brain's entanglements would be immediately regained. This fact explains additionally how people who are blind from birth would be able to experience vision in an NDE. The faculty of vision need not be based in the brain, but in the entanglements in the spirit beyond space-time. It also explains the seemingly spaceless and timeless nature of many of these states, where the subject felt that the ordinary categories of space and time could break down. Thus given this mentalistic framework, no contradiction exists between reports of the afterlife and the findings of modern neuroscience.

As well as reconciling the paradox involving the conflict between NDEs and the findings of neuroscience, the binary-soul doctrine reconciles seemingly conflicting evidence for different accounts of the afterlife. The two halves of the mind, spirit and soul have potentially different destinies after death. As knowledge of the binary-soul doctrine faded into obscurity over the ages, these two outcomes became largely separated. The outcome of the soul is found in the popular concepts of heaven and hell. Meanwhile, the understanding of the outcome of the spirit evolved into what has eventually become thought of as reincarnation.

Both beliefs regarding the afterlife have evidential support and it would be unwise and/or hasty to reject either simply because of one's religious background. For instance, there is the case Howard Storm, the atheist professor who went to hell and came back.[8] Likewise, there is the case of a Druse boy who remembered his past murder, correctly identified his murderer, and located his body.[9]

It would be unwise for a Hindu, for instance, to ignore warnings about hell, as Hinduism even has a concept of hell in the form of Naraka. In similar fashion, Svarga Loka has an obvious parallel to heaven. Likewise, it would be hasty for a Christian to discount accounts such as that of the Druse boy, as even John the Baptist was described as "the Elijah who was to come," and it was said that he was in the "*spirit of Elijah.*" The astute reader will recognize the reference to the term *spirit* rather than *soul*, as well the reference to John the Baptist as "another Elijah."

This reconciling of seemingly contradictory bodies of evidence, sometimes unexpectedly, is exactly what we would expect from a deeper account if it is true. The mentalistic account provides just such a way to reconcile both the conflict between neuroscience and near-death-experiences, as well as the conflict between varying accounts of the afterlife.

C. Psi Phenomena

As expected, The Principles of Mentalism and Correspondence can also account for a wide array of mental phenomena that would otherwise be considered "paranormal" or "supernatural." One of the most widely known of such phenomena is psi or so-called psychic ability. Though they could be considered "supernatural" or "paranormal" insofar as they go beyond materialistic science, they are nevertheless quite explicable in terms of the scientific framework given for Hermeticism before. First though, it is worthwhile to examine some of the evidence.

i. Evidence For Psi: The phenomena of psi is heavily contested, but very good evidence has been established for it. So-called psychic abilities of extrasensory perception such as telepathy, clairvoyance and precognition have been studied under laboratory conditions in a variety of laboratories. These include the Princeton Engineering Anomalies Research lab (PEAR), The Institute for Noetic Science (IONS), The Global Consciousness Project (GPC) and the Rhine Institute.

These studies have produced substantial statistically significant evidence for these phenomena. Thus based on the evidence alone their existence should not be in doubt to the honest mind. Some examples of this evidence is provided here.

1. One of the more significant experiments was conducted by the Dutch psychologist H.I.F.W. Brugmans in 1923. Here the experimenter determined one square at random from a 6 x 8 squared checkerboard, and the subject was asked to select the right square from behind a curtain. Furthermore, the subject's Galvanic skin response was tested to see if there was a difference between when he was about to select the right or wrong square. The experimenter would then try to tell the subject through his thoughts which square to select.

The results of this experiment were highly significant. Out of 197 trials just under a third of them, 60, were successful. The odds of this happening without the subject telepathically sensing the thoughts of the experimenter was calculated to be 121 trillion to 1.[10] In science the concept of standard deviation, denoted by the Greek letter sigma, or σ, is used to mathematically determine the level of scientific proof of something. Five sigma, which is the accepted standard for scientific proof, denotes a probability of 1 in 1,744,278. The results of this experiment far exceed this standard by seven orders of magnitude.

2. The "ESP Card Test" conducted by the Rhine Institute is another experimental test of psi with highly significant results. Here a deck of 25 cards was used. Each card had a symbol stamped on it with a total of five symbols, with five cards of each type. The deck was shuffled each time, and a person would select a card in secret and try to mentally send the symbol to a recipient. The recipient would then try to determine what symbol was selected.

Over the span of 188 such experiments the combined results were statistically highly significant.[11] To illustrate how significant this is, it would take 428,000 more tests averaging at the chance probability to cancel out the effects of the studies, providing statistically significant effects in favor of psi.[10]

3. The Pearce-Pratt distance telepathy test is another experiment which produced highly significant evidence for psi. This was similar to the ESP Card Test done by the Rhine Institute with 25-card decks. This test sampled 1,850 individual trials out of 74 tests and recorded 558 successful attempts. The expected number of positive attempts based on pure probability was nearly 200 below this though, placing the odds of the results not being caused by psi to be 1 in many billions.[12] This is again well above the 5 sigma standard and should already be sufficient scientific proof for the existence of psi.

ii. Skepticism of Psi: Given that ESP has already been long *established* as a real phenomenon by scientific standards, one may ask why it has not also been *accepted* by some members of the scientific community. After all even today much denialism can be seen on this subject with many ignorant skeptics of the phenomena pretending that no such evidence exists.

The answer to this is that something being *established as true* via *scientific proof* is different than that same thing being *accepted as true* by members of the *scientific community*. As is apparent from the evidence, the skeptics are either ignorant of the evidence or are prioritizing extraneous non-scientific opinions above the actual scientific evidence. If we take a closer look at some of this so-called "skepticism," the prioritization of these non-scientific opinions becomes obvious. A quote from a famous neuropsychologist Donald Hebb makes the source of his "skepticism" of psi obvious:

"Why do we not accept ESP as a psychological fact? Rhine has offered enough evidence to have

convinced us on almost any other issue... Personally, I do not accept ESP for a moment, because it does not make sense. My external criteria, both of physics and of physiology, say that ESP is not a fact despite the behavioral evidence that has been reported. I cannot see what other basis my colleagues have for rejecting it... Rhine may still turn out to be right, improbable as I think that is, and my own rejection of his view is - in the literal sense - prejudice."[13]

Firstly, it should be noted that this rejection of the evidence due to in Hebb's words "literal prejudice" does nothing to discredit the evidence. Rather the informed reader should recognize that it discredits the skeptics. For instance, if such thing as the skeptical JREF organization were credible in its skepticism, labs such as any of those who achieved the aforementioned results in studies would have already won the million dollars from the famed Randi Challenge. That they have not demonstrates to the intelligent observer the lack of credibility of such "skepticism."

Secondly though, the stated reason for Hebb's rejection of psi should be examined as it appears to also be the motivation behind the skepticism of many others. He stated that it was due to a supposed incompatibility of psi with physics and physiology. The issue with physiology was addressed earlier in the previous chapters and is no longer a problem due to quantum biology. The issues of physics deserve further exploration though.

iii. The Physics of Psi: The commonly supposed conflicts with physics come from relativity and quantum mechanics. Telepathy it is argued would send information faster than light, conflicting with relativity. It is also argued that in quantum mechanics entanglement can not be used transmit information and thus can not be used to explain telepathy.

Both of these criticisms would be valid if this is how telepathy worked, but these criticisms misunderstand how telepathy works. If we look at how these phenomena are explained with the aforementioned Hermetic-physics framework it becomes apparent that this is not the case.

According to the synthesis we have thus far derived, thoughts exist within the wave-function, which in turn exists outside of space-time behind the virtual reality. If systems in two people's brains become entangled then, their thoughts will share the same wave-function behind space-time, thereby forming an information channel between the two. Thus they could sense each other's thoughts via this shared channel, and the means by which they would be linked would not be physical, meaning it would not be within space-time. Similarly such things as precognition can be explained by entanglement across time. The supposed conflict with relativity is easily resolved once it is realized that the thoughts are not traveling through space but behind it. In Chapter II we briefly saw how the speed of light is a limitation that exists within the virtual reality as a constraint of how information processing manifests in it. However, the mental link is not within the virtual reality at all and can not be said to travel through space in any sense. It merely connects two minds in different locations of space without traveling through the intermediate space.

Likewise, the supposed conflict with the understanding of quantum entanglement is resolved once it is recognized what is meant by "information being transmitted by entanglement." What is meant by this is that *classical information* can not be transmitted from the entanglement, and this is in fact true. Any attempt to derive classical information from the wave-function involves measuring it, which would destroy the superposition and thus the entanglement.

By contrast in the case of telepathy, the thoughts being read are not classical information, nor is there any decoherence or collapse of the superposition at any point. The reason for this is that our thoughts are not classical information at all, but as demonstrated with quantum cognition, are within the wave-function in a state of superposition from the start. Like-wise the receiver of these thoughts is entangled with the transmitter. The information does not need to be extracted from the wave-function so as to be "read by" the receiver's mind, because the receiver's mind is already in superposition with the mind of the transmitter. Thus the quantum information is not destroyed upon collapse, as no such collapse occurs to get the information to begin with. Thus telepathy and related phenomena are quite compatible with physics as we know it.

The subtle assumptions the skeptics make of course are that information must travel through space-time and thereby be restricted by the speed of light. Likewise, they assume that the mind must be a classical non-quantum object and not non-local. Both of these assumptions are what one would expect if mind is not fundamental to reality and if nothing exists beyond space-time. Neither of these assumptions is necessary if one abandons a materialist framework of thinking though. The physics can work just fine without these assumptions. By contrast when it is put into another framework, such as the one given in the previous chapters, it explains these phenomena quite elegantly with no mystery.

D. Mental Phenomena of the Inner Space

This Hermetic-physics framework can also account for many other mental phenomena as well. A few of these will be explored here, along with their accompanying explanations. In each case it will be shown that there is nothing mysterious or mystical about these phenomena.

i. Prayer: The first phenomena we will look at is prayer. Here a person is said to communicate with God, and sometimes reports hearing things back. The efficacy of prayer is contested, and it may also be hindered by the low vibrational quality and wrong intentionality of the petitioner. Our focus here is not to look at the issues surrounding the efficacy of prayer, but rather to point out that a mechanism exists which can account for it that is readily explicable from a Hermetic perspective.

This mechanism is actually very simple. The Mind of The All, what religions refer to as God, generates the universe from the collective inner mental plane. In turn the petitioner talks to God in his or her own inner mental realm, which is of course a part of the greater collective inner mental plane. As The Mind of The All spans the entirety of this realm this Mind knows all and hears all within it. Likewise, from the perspective of the petitioner, if his or her vibration is tuned correctly, he or she will be able to receive information back from The Mind of The All. Though this subject of vibration will be explored more in the famous Principle of Vibration in the next chapter.

ii. Divination: The phenomena of divination can be explained by this framework as well. Specifically, it is closely related to the Principle of Correspondence and to the phenomena of synchronicity and the collective unconscious. The "As within, so without" formulation of the principle may be the best to understand this in terms of.

In Chapter III we learned how the world within physical space-time is a projection from a collective inner plane, what Jung referred to as the collective unconscious. Space-time emerges from quantum information in the wave-function, and once this fact is understood in light of quantum cognition, the "program" from which it is "simulated" is "written" in terms of archetypes in the collective unconscious. Thus archetypes within this collective unconscious correspond to patterns of events and images that manifest in the physical world. One need only identify these archetypes in order to be able to relate seemingly causally disconnected events or imagery. If one or several events or images appears in this pattern, then one could in principle "*divine*" the subsequent images and events from the pattern.

Many such patterns exist, and in principle everything in nature is a manifestation of some archetypal "programming" in the collective unconscious. As a consequence of this, many forms of divination also exist. To illustrate this point, it may be worthwhile to look at one well known such collection of archetypes in the collective unconscious to demonstrate an example of this.

In Chapter III the Hermetic Kabbalah was looked at. Specifically, it was noted that the Kabbalist Tree of Life is an overarching pattern within the collective unconscious which in turn simulates the virtual reality according to the Hermetic dictum "As Within, So Without." A famous archetypal pattern is associated with this, and specifically with the pathways between the sephirot on the Tree of Life. In addition to each pathway being associated with a Hebrew letter, each is also associated with one of the Major Arcana cards of the Tarot deck.

What happens then is that the subject feels to select one of the cards due to collective unconscious influences which pertain to some situation. Given that these collective unconscious influences are also the archetypes behind the events of the relevant situation, they may also relate to future events regarding that situation. Thus the subject can divine the future events by reading the cards. In essence what is happening is that the future of the "computer game" is being read by reading its underlying archetypal programming.

Mere knowledge of the mechanisms at work behind this phenomenon of course do not justify engaging in magickal practices. As the reader is well aware, knowledge is power, and power can be used or abused for good or bad ends. Our purpose is merely to demystify the phenomenon by providing knowledge of the mechanism behind it.

iii. Possession: The theme of demonic possession has existed throughout human history and predates modern religions. Rather than being a mythological phenomenon believed by superstitious people it has a basis in truth as occultists are well aware.

The framework of Hermetic physics also provides the possibility of a completely scientific account of it. As explained earlier, the realm physicists refer to as Hilbert space, mistakenly believing it only to be a mathematical tool, is actually what is known to occultists as the astral plane. In it exist wave-functions from which space-time emerges, and these wave-functions have an innately mental nature.

Given its mental nature, there is nothing inherently impossible in the physics for conscious agents to exist in this realm that are not tied to physical avatars within space-time. It is of course possible for them to have physical avatars in other space-times, that is to say on other vibrational planes, but this is the topic for the next section.

These beings would fit the description of what religion refers to as supernatural beings. Some of these beings are negative, and it is from these that legends of fallen angels and so on derive. These beings go by many names, demons, djinn, and archons are a few examples. Regardless of what they are called, they exist and are descriptions of the same thing, and have been able to influence humanity unaware for ages. This is perhaps the case more so in modern times due to the generally materialistic belief of the modern era, which they use to their advantage.

Existing in the collective inner plane, they are naturally able to influence the minds of humans also extended in that inner realm. As was explained earlier, the mind and its thoughts and processes are describable through quantum cognition as a wave of quantum probability. Thus all these beings need to do to influence a person is to try to get the person's thoughts to resonate at the vibration of the demon's. This allows the demon to influence the person's thoughts and behaviors through the principle of resonance. Sometimes this resonance becomes so strong that the person's personality, thoughts, and actions are taken over entirely, or at least substantially, in which case they are said to be possessed.

This can be seen even today in some modern musicians who describe themselves as having alter egos who influence their behavior in their performances. It should be noted that sometimes these personalities are noticeably different from that of the performer, such that it is apparent that a foreign consciousness has taken over. In other cases they will even mention that these alter egos have been "summoned up" and that they can not make them leave, illustrating the foreign nature of these personas and that they are not merely artifacts of personal psychology.

On a larger note, it should also be recognized that just as an individual can be possessed, so too can a large group of people be influenced in a similar manner, such that they can appear to be possessed. Here the concept of an egregore, or group identity comes into play. Many of these are in operation today, but many exist at subtler levels so as to control the behavior and beliefs of whole groups of people while masquerading under the guise of normalcy. The reader having been enlightened of their existence can break free of their spell.

E. Beings From Other Vibrations

Described also in the third chapter was the existence of vibrations of other degrees of freedom existing beneath space-time in the theory called Modified Newtonian Dynamics or MOND. These were paralleled with the vibrational planes referenced in *The Kybalion*. It is important to examine the physics of such planes, and as to how beings in such planes would appear to us, as a very interesting and curious phenomena is related to the existence of these planes.

First it is important to look at how beings existing in these planes would appear to us. As was noted before, these planes exist outside of our space-time and would thus from our perspective appear to be non-physical or what is commonly termed "supernatural." Thus the beings in them would likewise be defined from our perspective as "supernatural." Furthermore, given that these planes occupy the domain of Hilbert space, which per the science of quantum cognition is fundamentally mental, these planes could be seen from our perspective as being mental or "spiritual" in nature.

Thus from our perspective, the beings occupying these planes would by our standards be definable as "supernatural" and "spiritual" given our usual usages of these terms. Of course we have terms for beings that are "supernatural" and "spiritual." We call them demons if they are of a spiritually negative nature, and angels if they are of a spiritually positive nature.

What is peculiar though, is when we look at their physics from their own perspective. Given that collapse is relative to frame, everything will collapse its own wave-function from within its own frequency range. Schrodinger's Cat in the popular eponymous thought experiment is in a box and is therefore in superposition with respect to the experimenter. With respect to itself however, Schrodinger's Cat sees itself in only one defined state. The same principle holds here only for entire space-times being held in superposition with respect to us or not.

Thus curiously this property of being "outside of space-time" and "in the wave-function" and thus by extension "spiritual" as per what quantum cognition tells us about states in the wave-function if we take it to be more than a metaphor, is only relative to the frame of the observer. From our perspective beings in these other vibrational planes are "supernatural" and "spiritual." From their own perspective they would appear as the products of collapsed quantum states though, namely as physical beings.

Given that these planes would likewise obey the underlying laws of quantum mechanics and of the emergence of space-time, they would have properties similar to our own, only encoded in a slightly different frequency range. Naturally this would include central forces like gravity and electromagnetism, as well as the nuclear forces.

This would give rise to basic chemistry and biological evolution, as well as in the case of gravity, the existence of stars and planets. Thus they would see themselves as biological beings with stars and planets similar to our own. Furthermore, as will be explored in two sections, if any of them exist in what religion would refer to as a fallen state of a sufficient degree, they would need to rely upon technology as well.

These facts, as well as the curiously dual nature of these beings, helps to shed light on another phenomena that is often debated in modern times; the topic of UFOs and alien beings. Two chief opinions exist about these beings which are thought to conflict. On the one hand, there is the conventional view that they are merely visitors from other worlds. On the other, there is the view, that they are metaphysical beings, and are usually equated with the demonic.

Both of these views are simplistic, but it is not inaccurate to say that both are in the strict sense "true." It is true that in reports of these beings, they are said to possess technology, and they also appear to have a biological nature, indicative of a physical, material, and biological being. On the other, they are sometimes associated with paranormal activity, and some are associated with a malevolent spiritual influence. There is also the curious fact that sometimes UFOs are said to disappear in a manner similar to that of ghosts.

As we have seen here, these two seemingly contradictory accounts would both be true if a Hermetic account of the physics is understood. From our perspective, they would appear to be spiritual. From their own perspective as well as our own when they are phased into our own vibrational plane, they would be quite physical and no different from us in kind. This seeming paradox only appears from either a materialist account which leaves no room for the spiritual, or from a dualist account which reduces the spiritual to superstition and radically separates it from the physical. The account here reconciles both natures of the phenomena though.

It should also be noted that as observed in astronomy, the presence of dark matter closely matches that of real matter. This is seen in the fact that dark matter is strongly concentrated in galaxies where matter is present, and is very scarce in intergalactic space. If dark matter is indicative of the presence of these other planes however, then it may not be wrong to think that planes of a sufficiently close vibration should have stars and planets located approximately where they are in our own plane. They may in fact be thought of as the same stars in some cases, and existing in essentially the same universe were it not for the fact that their plane exists on another vibration.

Lastly, as the authors are both aware, some of the reports of these beings fit the descriptions of similar beings found in mythology and religion. These descriptions are not wrong, only limited in scope. They are not *merely* ancient aliens confused with demons or angelic beings, nor are they *exclusively* demons masquerading in a deceptive manner as aliens. The real deception is that this is some dichotomy rather than as the old parable goes, the descriptions of "people blind from birth when viewing an elephant." The reader should be able to see these same patterns when exploring these sorts of accounts in depth.

F. The Origin of the Pentagram

The pentagram has long been associated with occultism. It is traditionally said that it is associated with the five elements; earth, air, fire, water, and spirit or ether, with each element corresponding to one of the five points of the star. Spirit or ether is placed on top, symbolic of it ruling the material world. In the case of black magick it is inverted, with spirit placed below and ruled by the material world.

The meaning found in the symbolism associated with the pentagram may very well have Jungian significance, which in turn due to the Principle of Correspondence may have magickal or occult significance in its own right. Despite this, the pentagram is related to something else as well of occult significance. Specifically it is associated with the crossing between of planes; the above with the below or the within with the without for the purposes of summoning or banishing energies or even entities into and out of the physical plane.

The relation between the physical plane and the spiritual and mental planes was well defined on grounds of physics in the last chapter. This physics includes how things in the spiritual plane manifest as, or influence what occurs in the physical plane. Since the pentagram is used in rituals related to causing energies or entities to manifest from the spiritual and mental planes, one wonders if this might also show up in the physics as well. As a matter of fact it does.

Space-time emerges from quantum probabilities in the wave-function which define locations, which in turn causes space-time geometries to emerge from the underlying non-local reality. Prior to this point, space does not exist. It is only the locations collapsed from the wave-functions that define space. Any student of geometry will immediately recognize that space is defined by points and the connections between them. One-dimensional space will be defined by two points creating a line for example. Likewise, two-dimensional space is defined by drawing a third point and creating a plane, and so on. If we continue up like this, adding a special rule for time asymmetry, we can derive four-dimensional space-time and its contents, from underlying quantum probabilities in a similar fashion. In fact, this is exactly the approach that the quantum gravity theory of Causal Dynamical Triangulations, or CDT, specifies.[14] To do this CDT requires five points to triangulate a four-dimensional space-time, much as a triangle is needed to triangulate a two-dimensional plane. The four-dimensional object produced from these triangulations is called a pentatope.

However according to CDT, space-time is actually two-dimensional as seen from the Planck scale. Thus the four-dimensional geometries are really encoded in two-dimensions in the form of projections.[15] When a pentatope is translated from four-dimensions into its two-dimensional projection it produces a pentagram inscribed inside of a pentagon!

This is noteworthy as this pentagram is what transfers the information encoded in Hilbert space into our physical space-time. "Hilbert space" and "physical space-time" though are merely the scientific terms used to describe what the Hermeticists refer to as the spiritual and physical planes respectively. In turn the pentagram is what occultists use to summon energies or entities from the spiritual plane into the physical, mirroring what occurs in the triangulation process of CDT.

While this comparison is remarkable, the obvious objection will appear that the pentatopes of CDT are Planck-scale quantum objects existing in an objective environment and not the pentagrams drawn by occultists governed by classical physics. And this is quite right; mere chalk drawings existing in an environment are no more special than anything else, and like everything else in such an environment are governed by classical laws.

What is not realized though is that there is no objective environment! The collapse of the wave-function is relative to perspective, be that perspective a measuring device or a conscious observer. From the observer's perspective, the environment is in superposition itself and lacks defined reality before it is observed. When an observation is made, what is seen is not the environment itself but rather an *image or conscious perception* of the environment.

The chalk in the environment does not have any mystical properties, because the chalk, like the environment it is in, is not even real. Only the perception of it is real, and the perception does have quantum properties since it is the stream of data rendered from the wave-function, which is to say in Hermetic terms, from the mental or spiritual plane, or in Jungian terms from the collective unconscious. Once an environment, or rather the image of an environment, is collapsed into being then naturally any subsequent states in superposition will be very tiny indeed. The space-time would need to be triangulated from these at a subatomic scale. However, the space-time of the initial collapse of the environment can only be triangulated into being from the perceptions of the observer. While everything one perceives does this, the point of the pentagram is to draw the attention of the mind of the occultist. Now what is the image of the pentagram produced from prior to collapse? It is the wave-function in Hilbert space of course, which as we saw in Chapter III is really the collective inner mental plane. This collective inner plane is where the energies and entities are summoned from.

Thus to cause an entity or energy to manifest using the pentagram to bridge the inner and outer worlds, the practitioner invokes what he or she is seeking to manifest. The invocation one may say is merely air waves, and this is true. Just like the chalk used to draw the pentagram, the air waves are merely classical objects within the virtual environment. They have no special properties in themselves, just as quantum mechanics tells us they have no intrinsic reality either. However, the sensation or auditory "image" of them on the practitioner's consciousness is very real and is used as a means to activate or tap certain things in the collective unconscious. As this is same collective unconscious that is being bridged by the *image* of the pentagram in the first place, it allows what it is being invoked to manifest into the physical plane via this bridge.

The reader should be warned though, that attempting to control the goings on of the spiritual world is a direct inversion in the order of nature. There are severe consequences for the practice of black magic as the Principle of Cause and Effect would suggest. We are only showing how this works here to illustrate the reality of such practices, and how they are not at all mystical or beyond comprehension.

G. Fallen State Metaphysics

Though the "fallen state" as a concept is often associated only with religion, it can be accounted for in terms of occult principles as well. These principles are chiefly Mentalism and Correspondence, though as will be seen, the Principle of Vibration is also relevant. The Principle of Gender seen in Chapter X also comes into play. These principles as we have seen before can be derived purely in scientific terms with our most advanced physics and are not religious or based on dogmatic belief in any way.

The occult understanding of the fallen state is also directly relevant for another topic in the study of occult science, that is regarding the origin and nature of magical languages such as are used in spells. This will be looked at in the next and last section of this chapter. For now, we will examine the fallen state in terms of Hermetic principles. Along the way it will become apparent that this concept is not contained to Western religion but can be found under different guises in Eastern religion as well, in addition to the esoteric traditions of Hermeticism and Kabbalah.

i. Occult Science of The Fallen State: The concept of the fallen state in religion is that the world or at least the human condition was once free of both external evil in the form of pain and suffering as well as from internal evil in the form of human vices. This is said to have changed during the "fall of man" into a state which has both external suffering and inner evil in man.

This direct correlation between inner evil and outer evil is an immediate clue that the fall has a connection to Hermetic science. The "As Within, So Without" formulation of Correspondence comes to mind. The outer evil of suffering and natural evil mirrors the inner evil in the form of vice and ill-will. Given what we have already learned from Mentalism and Correspondence, we are in a good position to begin understanding the fallen state from an occult perspective.

To understand evil we must first understand good. In the Divine Pymander, Hermes stated that:

"Good, O Asclepius, is in none else save in God alone; nay, rather, Good is God Himself eternally."

God of course is The Mind of The All understood by Mentalism in Chapter II. The Mind of The All is good out of necessity, as it is the originator of all other conceptions of good in all other emanated minds. Thus no perspective in any lesser mind could disagree with the Goodness of The Mind of The All, *if it were sufficiently enlightened.* As a result, God being good only creates or emanates what is good.

Evil must therefore be parasitic on Good, existing as a contingent thing. For there to be evil there must first be good. Applying this understanding to "As Within, So Without" we immediately discover something very recognizable about the fallen state.

States in the inner space exist as *knowledge*, whereas their corresponding states in the outer space are *things in themselves*. Thus if we are to ask where evil arises from in the world, it must come from the source of the simulation, the inner space and the states of *knowledge* within it. Thus physical evil must be the product of knowledge of evil in the collective inner mental plane. As evil can not exist without good, we further see that the existence of evil in the world is due to the *Knowledge of Good and Evil* being had in this inner plane. This is immediately obvious in its connection to the Garden of Eden story wherein man caused the fall by knowing the Knowledge of Good and Evil.

If we extend this investigation further, we discover many other seemingly disparate features of the fallen state all arising from the same source. Firstly, if we ask what evil is, we find that if it is not an emanation from The All, it must be a lack. But things can not go missing from The Totality. The Totality remains The Totality, and thus by its very nature, nothing can go missing from all that is. Thus it must be a relative lack of a single thing missing some other thing, the product of a distortion in the good where something good has been broken apart such that it is missing a part of what makes it good.

What is broken though to produce this evil state? As per Mentalism it is states of consciousness that are broken. This is what physicist Tom Campbell refers to in terms of entropic consciousness in *My Big TOE*. This is not always apparent in the physical world. A broken window is bad, but so is a murderer who apparently does not have any broken pieces. As per Correspondence however we know that the evil in the outer world is the product of evil in the inner world. Thus it is consciousness broken down in the inner mental world. These inner planes are the same as the emanating world that the physical plane comes from, which is veiled from the physical senses.

This replicates yet another aspect of the fall described from occult science, that of the Kabbalist *Shattering of the Vessels*. As we saw with the derivation of Hermetic Kabbalah in Chapter III, Hermeticism and Kabbalah can mesh together very neatly. The emanating world of Kabbalah is the inner world of Hermeticism, and according to Kabbalah we see the physical world as a kind of veil, concealing the true emanating world. This neatly matches how the French physicist Bernard D'Espagnat describes the quantum reality as a "veiled reality[16]" from which our senses only see the other side of the veil.

The fall on Kabbalah is described within this emanating world in terms of broken emanations from God called Qlippoth. The event which brought this about is referred to as the Shattering of the Vessels, and describes the shattering of consciousness in the veiled reality of the emanating world. This schism in consciousness can alternately be thought of in wholly scientific terms as broken states of entanglement in the Hilbert space beyond space-time.

Given this framework, the reader can see how the contrasting unfallen state is absent evil either of the inner variety or the outer variety. In a perfectly unfallen state all minds would be entangled with each other in an unbroken fashion. In such a state, all such minds would automatically share introspective states through entanglement, and thus this unfallen state would correspond to a high degree of psi in the form of telepathy. As a consequence, this would cause everyone to experience the pains and pleasures of everyone else. Thus there would be no desire to inflict such pains in the form of human evil, as it would be a pain inflicted against oneself as well. Likewise all would seek the good of each other as all would experience this good through this collective introspective link.

Likewise when it comes to natural evil, the link of each mind to The Mind of The All, the spiritual connection to God, would account for the absence or natural evil. As a mind would have a connection to the Mind behind the simulation, the needs of the lesser mind could influence the environment of the simulation in a manner akin to telekinesis. This would in turn eliminate the uncontrolled natural elements that create natural evil within the mental construction of physical reality.

It should be stated, as will be discussed later in this section, that the fallen state occurs in degrees. Thus it need not be the case that here on earth there was no biological death or eradication of all illness before the fall in contradiction with the paleontological record. There was however a greater degree of spiritual connection and with it a greater mental influence over the natural elements, resulting in less of a need to rely upon tools, technology and the like. All this will be discussed shortly. First though, with this occult understanding of the fallen state, we will now examine it in light of comparative religion.

ii. The Fallen State in Comparative Religion: Though the doctrine of the fallen state is most prevalent in the Abrahamic religions it exists in other forms in all main religions. These include both Eastern and esoteric beliefs. We will include these here, including the just discussed esoteric systems for completion.

1. Abrahamic Religions: Common to all three Abrahamic belief systems, Christianity, Islam, and Judaism, is the belief in the fall of man, and it is from this tradition that the fallen state is primarily known. This fall is caused by the Knowledge of Good and Evil and resulted in both inner evil, sometimes called "Original Sin," as well as natural evil, often called "The Curse." The explanation for this was just discussed, and it is not necessary to repeat here. It is worthwhile to note though the emphasis on forgiveness in these traditions, as forgiveness is what reunites the broken fragments of consciousness comprising the fallen state, as well as reconnecting these broken pieces with God.

2. Eastern Religions: Hinduism and Buddhism are the two primary examples here. Though connected, Hinduism also has a concept of *Maya*, which is worth examination in its own right.

-*Hinduism:* The concepts of duality and Maya are relevant here. Duality is the state of consciousness in separation. Here minds treat each other as different rather than united in a state of *advaita* or non-duality. This is merely another description of the fractured entanglements comprising the fallen state discussed earlier. While in this state, the minds are under Maya or the "veil of illusion," referring to the illusion of the material world. This is also a logical consequence of the fallen state, as the fallen minds no longer have introspective access to The Mind of The All that generates the physical world. Lacking this connection to the Mind behind the simulation, they also lack the direct awareness that would go with such a connection, that the world is a mental construct rather than an independent material reality.

-*Buddhism:* Buddhism has the concept of ego consciousness, and the goal of Buddhism is to transcend this state of ego into enlightenment. Ego consciousness refers to consciousness in separation, again paralleling the fallen state as described in terms of fractured consciousness.

3. Esoteric Traditions: These were discussed, but it is worth summarizing them again. Additionally Hermeticism has another concept related to Vibration that is linked with the fallen state.

-*Kaballah:* The fallen state in Kabballah is described as a cosmic catastrophe in the emanating world behind the veil of the physical world. This is referred to as the *Shevirat haKeilim* or Shattering of the Vessels, wherein the various emanations from God, called Sephirot, were broken into shattered fragments called Qlippoth, which translates into "shells" or "husks." It is perhaps these shells or husks that Jesus referred to as "the chaff."

   -*Hermeticism:* In addition to the Hermetic understanding of the fallen state discussed earlier, Hermeticism has another unique description of the fallen state, not seen in the other traditions. This is found in relation to the Principle of Vibration, which we will explore in more depth in the next chapter. In Hermeticism the fallen state is associated with a drop in vibration whereas an unfallen state, or perhaps a less fallen state, corresponds to raised vibrations. This is due to the rate of vibration of wave-functions increasing when there are more states or energy contained within them, and decreasing when there are fewer. The entangled state of the unfallen state, having more consciousness agents within it, thus has a higher vibration than the unfallen disentangled state. Though the Hermetic origins of this are not popularly known, this concept has found its way into modern New Age thinking.

  Before moving on it is worth noting that the escape from the fallen state can only be achieved through the help of The Mind of The All. Only God is perfectly beyond ego or separation, and thus only God can release one from a state of separation. One can not do it oneself anymore than one can open a door by at the same time pushing against it with one's feet to gain leverage. This is precisely because all motivations to escape ego arise from within ego. Likewise, no finite priests, bodhisattvas, angels, gurus, gods, guides, or religious institutions can accomplish this for the same reason. All other such focuses are idolatry.

  iii. Degrees of the Fallen State: Lastly, it should be noted that though the fallen state is a distinctive condition, this condition comes in degrees. This accounts for the fact that death was apparent in the paleontological record long before the fall of man on earth. It is worth noting as well that unlike with some assumptions, the fall of man came some time after the inception of man as a species in this world. As we just learned, the fallen state corresponds to the conscious elements in the collective inner plane being broken apart and disordered. Thus the degree to which they are disordered represents the degree to which this fallen state is fallen. As a result, some states will be substantially more fallen than others. In some cases, one could live in a relatively unfallen state such that one has a spiritual connection with God as well as with others. However, it is still possible for such a state to be not sufficiently ordered to such a degree that it corresponds to a physical reality which eliminates biological death and disease entirely. Nevertheless, it would not be spiritually dead, and the mind would have ready access to the collective inner plane. Thus upon biological death their souls or kas would not be disassociated from their spirit or ba, what the ancient Egyptians referred to as the "second death,"[17] a concept seen later in the Christian religion. This was the state prior to the event known as the "fall of man."

  This event was the result of a race of fallen beings coming from elsewhere to tempt man to "become as gods" themselves, thereby breaking their consciousness off into a state of ego with nothing higher than it. These beings are known variously in legends throughout the world as the nachash, the chitauri, the nagas, and so forth, all of which are associated with the serpent, as students of mythology can confirm for themselves. The existence of these beings is not merely a metaphor either, as these myths have a basis in actual facts, as advanced occultists are well aware.

  One may ask why the state prior to the fall of man was not perfect. This is a valid question and it has in part to do with the Principle of Vibration, which we will explore shortly. The degree to which the broken pieces of the fallen state are re-ordered and re-entangled corresponds to their vibration, with higher vibration corresponding to a lesser degree of fallenness.

  The reason is that this particular vibratory plane is not at this state. Other such vibratory planes exist as discussed in Section E of this chapter and in Section A of the previous chapter. Many of these are of a high enough vibration that they would be "perfect worlds" as we would imagine it, without death and so forth, and with a far greater degree of mind-dependency. It is from one of these planes that the so-called fallen angels originally fell into this plane. Being in this plane though, we do not have the same kind of perfection we would see in a heavenly plane.

H. The Origins of Magical Language

Having explored the fallen state, and by extension the nature of the pre-fallen state, we are now in a good position to explore the topic of magical language. As explained before, prior to the pre-fallen state the state of consciousness was such that minds could communicate with other minds through means of telepathy. As a result, speech was not necessary. Though this certainly does not mean it was not used before, it was not necessary to such a degree that widely established languages were necessary.

Thus the existence of widely established languages is a phenomenon that would have appeared after the fall. Prior to this, there would have been a sort of "mental language" or "mentalese" common to the structure of thought itself through which people could communicate mentally. Such ideas are already developed by modern linguists such as Jerry Fodor in his "Language Of Thought Hypothesis" or LOTH.[18] This mentalese would be very basic and would break down into individual bits of sensation. In the case of sound or auditory sensation, this would be expressed as basic phonemes, individual sounds comprised of single vowels and consonants.

The earliest languages then would simply be the external verbalization of such "mentalese." And in fact when we look at the earliest languages such as Sumerian, Hebrew, or Sanskrit we find that they are agglutinative. Sumerian words for instance are comprised of many simple syllables representing a simple phoneme each. In turn, each of these syllables has an individual meaning that exists either independently from or in conjunction with the other syllables it is joined to.

As these languages are verbal representations of the simple phonemes and auditory sensations naturally encoded into the wave-function, aka the collective inner mental plane, it can also be said that the physical plane is in some sense "programmed" by them. Given that the control of such programming is how the principle behind magick works, this forms the origin of the concept of magical language. For example, in addition to being a spoken language, ancient Sumerian was said to possess magical properties useful in the creation of incantations and spells. Likewise, in the Kabbalist tradition the universe is said to be an illusion continuously generated by God through the divine speech of Hebrew letters. In fact, the very term "abracadabra" derives from the Hebrew for "I create what I speak."

It should be noted that while spells can work, the acoustic sound waves do not have any magical effect themselves. Rather their conscious perception influences the inner realm of the magician, enabling him to input such a spell or "code" into the inner realm, thereby reprogramming the outer world.

Many other phenomena associated with Mentalism and Correspondence exist, and are worth exploring in their own right. However, the ones here serve as a good introduction. For sake of brevity though, we will now move on to the fascinating subject of vibration of which many links to modern physics exist.

V The Principle of Vibration

The Principle of Vibration is perhaps the third most important principle in Hermeticism. This principle is both simple yet very significant as it has very strong and easily identifiable connections to modern physics. In turn from the physics of both this principle and the previous two, the remaining four principles can be derived fairly simply. First let us examine the principle itself though from *The Kybalion*:

"The great Third Hermetic Principle–the Principle of Vibration–embodies the truth that Motion is manifest in everything in the Universe–that nothing is at rest–that everything moves, vibrates, and circles."

This tells us that everything vibrates. Indeed, in physics we find that everything does vibrate, be it in terms of light waves, sound waves, or molecular vibrations and Brownian motion. In fact, it is common for those studying Hermeticism to believe that this is what the Principle of Vibration is speaking of. Though all of these things are indeed various products of the underlying vibrations referred to in Hermeticism this is not what the Principle of Vibration is referring to as vibrations.

When we look at how these vibrations are described consistently in *The Kybalion*, and recognize the related physics behind the principles of Mentalism and Correspondence, one thing springs to mind. These vibrations are the uniquely quantum vibrations described as the quantum probability waves in Chapter II. Once Vibration is examined in this light it becomes clear that everything it is referring to with the concept of vibration meshes perfectly with the findings of modern physics. Some of these discoveries are quite advanced as well and have only been discovered in the 21st century.

A. Quantum Probability Waves

As it turns out the concept of vibration is central to our most fundamental physics in the realm of quantum mechanics. As described in Chapter I, particles have a wave-like nature when not observed. This wave-like nature is referred to as a quantum probability wave and is described mathematically with the wave-function, which is provided to us in the Schrodinger equation[1] as the Greek letter psi or Ψ:

$$\frac{d^2\Psi}{dx^2} = -\frac{8\pi^2 m}{h^2}(E-V)\Psi$$

This is a kind of differential equation known as a wave equation. When it is solved for any particular system, it provides a wave-function that describes the particular system. This wave-function comes in the form of a waveform which has an associated *vibration* of k.[2] Given that k is dependent upon the energy of the system and that everything possesses energy, it follows that everything has a vibration. This is the first obvious parallel to Hermeticism:

"Nothing rests; everything moves; everything vibrates."

If we continue we find that these parallels go beyond this superficial level. The vibration rate or k-value of the wave-function is dependent upon energy, with higher energy corresponding to a higher rate of vibration and lower energy corresponding to a lower rate of vibration. This is paralleled again in *The*

Kybalion in its description of the increase of vibration as an object moves upwards in energy in the direction of merging with The All:

"Science does not dare to follow the illustration further, but the Hermetists teach that if the vibrations be continually increased the object would mount up the successive states of manifestation and would in turn manifest the various mental stages, and then on Spiritward, until it would finally re-enter THE ALL, which is Absolute Spirit."

The closer an object gets to The All, the greater portion of The All's energy it would possess, and thus according to the Schrodinger equation the greater the rate of vibration its wave-function would possess. This is neatly paralleled here in exactly the same way minus the modern mathematical language of quantum mechanics. Going on, we find yet another parallel:

"The Hermetic Teachings are that not only is everything in constant movement and vibration, but that the "differences" between the various manifestations of the universal power are due entirely to the varying rate and mode of vibrations."

This is paralleled in the Schrodinger equation again. Every material particle or collection of particles fundamentally reduces to energy, and the only thing that distinguishes wave-functions from one another is the form and degree of their energy content. This energy content as described before correlates directly to the vibratory rate of the wave-function. Ultimately everything boils down into energy quanta and thus differences between things reduce to differences in energy quanta, which have associated vibrations.

Of course a fast moving proton and a slightly slower neutron can have the same energy value and yet have different properties. The proton has a positive charge, while the neutron is uncharged, and the two have differences in quark composition as well. Thus it appears that the quantum description only grasps this particular Hermetic concept regarding vibration indirectly.

Perhaps a better piece of physics that does grasp it is superstring theory. In string theory, the differences between particles are found only in the differences of vibrations of superstrings. This also supports the above statement from *The Kybalion*, albeit more directly. Of course these differences also boil down to differences of energy, and these superstrings have their own wave-functions as well, but the connection is more apparent when string theory is brought in to explain these.

The uniquely quantum nature of these vibrations can be seen more explicitly though in the next two statements. The first regards the mental nature of vibration, and the second the non-local nature of vibration.

"Every thought, emotion or mental state has its corresponding rate and mode of vibration."

In Chapter II it was demonstrated that the quantum probability waves of quantum mechanics were explicitly immaterial. In fact, the most basic facts of quantum theory are flatly incompatible with materialism in general. Material objects have physical properties before we look at them. Quantum objects do not. In addition to being immaterial, these waves were described as explicitly mental via quantum cognition, also as noted in Chapter II. Thus "every thought, emotion or mental state" would have a "mode of vibration" specified by the wave-function encoding it.

These vibrations are, for the most of the time in the case of the human mind, built into quantum biological structures in the brain. Though as we will see now, they can be non-local as well:

"But the Hermetic Teachings go much further than do those of modern science. They teach that all manifestation of thought, emotion, reason, will or desire, or any mental state or condition, are accompanied by vibrations, a portion of which are thrown off and which tend to affect the minds of other persons by "induction." This is the principle which produces the phenomena of "telepathy"; mental influence, and other forms of the action and power of mind over mind,"

Here *The Kybalion* describes something that can be uniquely and only quantum mechanical in nature. Ordinary vibratory waves dissipate radially from their source such as with sound sources or radio waves dissipating as one gets far enough from the radio station. Furthermore, the classical waves produced in the brain are far too weak to travel far.

Thus what is being described here with telepathy is the sort of "spooky-action-at-a-distance" Einstein used to describe quantum entanglement. Of note are the facts that *The Kybalion* is describing the "at-a-distance" phenomenon of telepathy, and that it is associating this phenomenon with a vibration. The only kind of vibration that hosts at-a-distance phenomenon are quantum waves in entanglement.

The parallels between the Principle of Vibration with quantum mechanics are remarkable, but we are not done yet. Even more striking parallels are yet to come.

B. Vibrations of Space-Time

You may think that the Principle of Vibrations, or for that matter, the wave-nature in quantum mechanics, applies only to particles of matter, but you would be wrong. Quantum mechanics applies to the fabric of space-time as well. This is also reflected in the Principle of Vibration as it is described in *The Kybalion*.

"The Universal Ether, which is postulated by science without its nature being understood clearly, is held by the Hermetists to be but a higher manifestation of that which is erroneously called matter–that is to say, Matter at a higher degree of vibration–and is called by them "The Ethereal Substance.""

The reference to the ether is an historical artifact of the science of the day, as the classical concept of an ether pervading space was still popular at the time of *The Kybalion's* writing in the early 20th century. Einstein's Theory of Special Relativity had only been published three short years earlier in 1905 and was not yet well known. However the concept of empty space or even the relativistic fabric of space-time having an associated vibration has a strong foundation in modern physics.

First there is the famed Zero-Point Field stretching across "empty" space. This field is the result of the Heisenberg Uncertainty Principle, which states that no system can have a completely defined energy value, being applied to empty space. Empty space would otherwise violate the uncertainty principle, as it would have a precise energy value of zero. Because of the uncertainty principle it does not though. Rather there exists a vibrating field of quantum flux, across the vacuum of "empty" space.

The parallels do not stop there though, even more interesting and illuminating parallels appear in research connecting space-time to the phenomenon of quantum entanglement. This link between space-time and entanglement was discussed in Chapter II in greater detail, but we will return to it again briefly.

In 2004, Xiao-Gang Wen and Michael Levin discovered that space-time behaves as a vast network of entanglements, described collectively as a string-net liquid.[3] Thus space-time itself would have an associated quantum vibration. This neatly matches the research discussed earlier, which has space-time emerging from quantum states also.[4] In 2015, Hirosi Ooguri of the University of Tokyo followed this research and successfully described space-time as being built up from quantum entanglement.[5] Quantum gravity tells us that the fabric of space-time itself has a vibration!

Details of Wen and Levin's discovery[6] shed light on another interesting part of the statement from the Kybalion. The substance of the "ether," which in modern terminology we know to refer to as space-time, is actually the same thing as matter at a different state of vibration:

"Matter at a higher degree of vibration–and is called by them "The Ethereal Substance.""

The *New Scientist* article on their discovery states that:[7]

"They also found that string-net theory naturally gave rise to other elementary particles, such as quarks, which make up protons and neutrons, and the particles responsible for some of the fundamental forces, such as gluons and the W and Z bosons."

Here the vibrations in the underlying string-net liquid, the network of quantum vibrations previously associated with the fabric of space-time, gives rise to the subatomic particles as well. In other words, matter and "the Ethereal Substance" are both products of the same kind of "vibration" and are fundamentally the same in nature. The only difference between them is their rate of vibration, just as *The Kybalion* tells us.

This neatly parallels string theory which tells us the same thing in a slightly different way. Particles such as the graviton, which are nothing more than bits of quantized warped space-time according to general relativity, are superstrings at particular vibrations. Other particles are also superstrings at different vibrations. Thus both space-time and matter are unified as merely different expressions of superstrings vibrating at different rates.

These parallels to quantum gravity, some of which was not yet discovered until recently in the 21st century are fascinating, given that *The Kybalion* was written in 1908 and claims to be based on even earlier sources. We are *still* not done though. Yet another distinctive parallel exists between the discoveries in the various fields of science associated with quantum theory and Hermetic teachings.

C. The Vibration of The All

In the chapter on Vibration, The Kybalion also describes the vibration of The All. This too has a precise connection to modern scientific knowledge *discovered fifty-nine years after The Kybalion was written*:

"Not only this, but that even THE ALL, in itself, manifests a constant vibration of such an infinite degree of intensity and rapid motion that it may be practically considered as at rest,"

This description of the vibration of The All is somewhat confusing and seemingly contradictory at first glance. It is uniquely described as having both an infinite vibration and no vibration at all such that it could be considered to be "at rest." When one considers what modern physics says though this all makes sense.

As something increases in energy the vibration of its wave-function will increase. Thus if one were to have an infinite amount of energy, as one might naively think The All to possess, it should have an infinite degree of vibration as The Kybalion describes. But not everything is at it seems.

The universal wave-function was first described by John Archibald Wheeler and Bryce DeWitt in 1967 in a variation of the previously stated Schrodinger Equation known as the Wheeler-DeWitt equation which we give here:[8]

$$\frac{d^2\Psi}{dx^2} = -\frac{8\pi^2 m}{h^2}(0)\Psi \text{ or } \frac{d^2\Psi}{dx^2} = 0 \text{ or } H|\Psi\rangle = 0$$

As the reader will note, the net energy value or Hamiltonian, H, that the Wheeler-DeWitt equation derived the universe as having, is zero. The reader may be confused as to how the entire universe filled with innumerable galaxies could have zero energy. This is due to its having equal amounts of negative and positive energy, canceling out at exactly zero. This is not mere theory either, but has been corroborated by cosmological studies of the net curvature of space-time as seen in astronomical data.[9] Zero energy also corresponds to a wave-function with a k-value of zero, producing no vibration at all. The All, as The Kybalion would put it, *"may be practically considered as at rest."* The Hermetic teachings on vibration again precisely match those given to us by modern physics.

What is interesting to note about the cases given in this chapter is that the teachings on Vibration include direct parallels with physics discovered only *after the fact*. This includes the general discovery of quantum mechanics and its associated quantum vibrations with the Schrodinger equation (1926),[10] the non-local nature of these vibrations (1935),[11] the zero vibratory rate of the universal wave-function (1967),[12] the discovery of space-time as a network of entanglement with which matter is also unified (2004 & 2015),[13,14] and the mental nature of these vibrations (2015).[15] *The Kybalion* by contrast was written decades, and in in some cases a full century before in 1908.

This fact is additionally corroborated by how the writers of *The Kybalion* did not themselves have a full grasp of science behind the Principle of Vibration. This can be seen in their reference to the ether, as well as to their assumption that the vibrations of the Principle of Vibration were referring primarily to the classical light, sound waves, and molecular vibrations known in that time. The fact that the manipulation of vibrations is associated with the production of miracles, as will be seen in the next chapter only further supports this. Microwaves and radios for instance operate by means of the manipulation of classical vibrations, but we would scarcely call these machines miraculous.

In hindsight, we see that the principles *The Kybalion* was dealing with paralleling modern physics are bizarrely anachronistic and prescient for the state of science in 1908. What is even stranger is that the authors of *The Kybalion* claimed to have gotten these principles from a much more ancient tradition. This supports the general thesis of the authors that the occult science of Hermeticism is both real, and is based upon a lost scientific understanding given to mankind at some point in antiquity.

The next Hermetic Principle, the Principle of Polarity, only expands upon these already discovered parallels with modern science. We will look at the Principle of Polarity in time. Before that, the next chapter will explore some of the phenomena associated with the Principle of Vibration.

VI The Phenomena of Vibration

Alongside mentalism and correspondence, vibration is one of the most important Hermetic principles. Many phenomena are directly linked with it. This should be expected, as according to both the principle itself and the physics which backs it up, literally everything has a vibration. Thus as one would naturally expect, both vibration and the phenomena associated with it are ubiquitous. Given its importance, a separate chapter has been devoted to examining some of those phenomena. Most notable among these is the mechanism behind miraculous effects or magickal powers, which we will now explore.

A. The Mechanism of Magick
The mechanism behind magickal or so-called supernatural powers is directly tied to the Principle of Vibration. This is to be expected as the physical state of a system both prior to and after a miraculous event will both include an associated wave-function with a specific vibration to describe it. Thus a miraculous occurrence can quite literally be described as a shift from one set of these quantum vibrations to another. In fact, this is precisely how *The Kybalion* describes miracles:

"A little reflection on what we have said will show the student that the Principle of Vibration underlies the wonderful phenomena of the power manifested by the Masters and Adepts, who are able to apparently set aside the Laws of Nature, but who, in reality, are simply using one law against another; one principle against others; and who accomplish their results by changing the vibrations of material objects, or forms of energy, and thus perform what are commonly called "miracles."

The vibrations being referred to here, whether the actual authors of *The Kybalion* realized it or not, are not the common mechanical or electromagnetic vibrations described by classical physics. We have machines that change these kinds of vibrations all the time, but we do not consider them to be miraculous. For example, a microwave changes the vibrations of water molecules in food, thereby raising their temperature from cold to hot, but this is hardly considered to be miraculous.

The case of quantum vibrations is different though. Firstly, the physics they relate to has intrinsic connections to Hermetic occult science as explained in this book. Secondly, as explained in Chapter III, the quantum states in the wave-function are not actually part of the physical world, that is the world inside of space and time. Entanglement reveals that these quantum states must exist outside of the virtual world we call physical space, lest they violate the speed of light limit imposed by special relativity. This realm outside of space-time is what is commonly described as the supernatural world though. The nature of this realm was discussed in some detail in Chapters III and IV. Thus the manipulation of these vibrations, or what are projected as vibrations within space-time, is in reality a manipulation of the quantum states existing within the deeper level of reality called "Hilbert space" in physics, and in common terminology "the supernatural realm." If one is changing the state of affairs in the physical world through "supernatural" means though, this is commonly called a "miracle."

The means by which change in vibrations cause this is simple enough to understand. Each wave-function, or "vibration" as Hermeticism would call it, encodes a set of states superimposed in various ways as quantum information. This quantum information is of course the "program" from which the physical reality in space-time is programmed. If one changes the underlying vibrations of these wave-functions, the states contained within them superimpose on one another differently. This creates different qubits, with different probability distributions of their associated wave-functions. Upon observation, these wave-functions would spontaneously produce entirely different physical states than what they were previously though, due to the change in "programming" underlying them.

Similarly, if we lived in a computer game unaware, such as in The Sims, and someone within it were to hack the underlying programming, they would also change reality. If we lived in such a world, not realizing it was virtual, it would seem to violate the laws of physics of that world insofar as they were programmed in at a surface level. Not realizing that these laws are the product of a deeper layer of programming, which itself could be changed, we too would call such an effect a "miracle" or "magic." The physics of this works in reality too. Of course it has been largely rejected by so-called "skeptics." However, as we have seen earlier, these sorts of individuals have also dismissed scientifically validated psi results far exceeding the five sigma requirement of scientific proof as well. Thus their judgement and authority in such matters is doubtful and very much in question.

Dr. Masaru Emoto's experiments with water illustrate just such an effect. His experiments discovered that water crystals could either become ordered and beautiful or disorganized and ugly depending on what mental intentions were sent to them. In one experiment conducted by Emoto, Dean Radin, Gail Hayssen, & Takashige Kizu, 2,000 people focused positive intentions on water samples.[1] These water samples were crystalized and compared with water crystals frozen from a control group. The differences between these two groups was such that it supported the thesis that mental intent could change how water crystals formed and was consistent with similar studies mentioned in the study.

The reader should be reminded here that if we are to take quantum cognition as literally true rather than as a metaphor, then mental intent is a quantum state in Hilbert space. Thus it is natural to assume that mental intentions could change the quantum state of any water molecules they may be entangled with, and thereby change their physical appearance as well.

In fact it is this change of mental states that the concepts of transmutation and alchemy were originally meant to refer to. During the middle ages when these concepts were driven underground, they were disguised as a form of chemistry seeking to turn base metals into gold. What they were in fact speaking of though was the mental transmutation of the soul into a refined state. This is a topic for elsewhere though. For now we will examine another subject raised by the experiment with water crystals, the relationship between geometry and vibration.

B. The Geometry of Vibration

If the geometric appearance of water crystals can change due to changes in underlying vibratory patterns sent to them, then it would follow that the principle of vibration would be associated with a great many other geometric phenomena as well. In fact this is seen all the time in the field of cymatics. Distinctive repeating geometric patterns appear as the result of the application of vibratory frequency. The result of this is that many phenomena are associated with vibration.

In the case of quantum vibrations, they are representations of the quantum states existing beneath space-time, which if quantum cognition is taken into account also exist as mental states. Given that these quantum vibrations are our window into the occulted world behind space-time, it is natural that some of these phenomena should have a relationship to occultism. We will briefly explore a few of these here.

i. Sacred Geometry: The topic of sacred geometry is popularly known among the public and ascribes religious or spiritual significance to geometric patterns. If we were to break this down into the scientific framework given in this book, the category "spiritual" would have to fall under the category "mental" or perhaps "cognitive." Meaning more precisely, sacred geometry is the topic relating consciousness to geometry.

This is to be expected as vibration has a direct correlation to mental states as per our previous discussion of quantum cognition. Thus so too should mental states be correlated to the various geometries produced by complex patterns of these vibrations.

The topic of sacred geometry is far too vast for a comprehensive treatment here. Furthermore, not all of it has been sufficiently integrated with the cymatic geometries found in mathematics to yet properly account for in a scientific fashion. Nevertheless, some of it has, and some of it also displays significant hints as to how it could be integrated into a scientific framework for future mathematical investigation. One example to explored here is that of the Platonic solids.

The Platonic solids can be derived very simply from the conditions placed on the wave-function itself. The wave-function can not have any singularities or discontinuities in its form, and it must also match up with other wave-forms or come to zero at its boundaries. This is true not just of quantum mechanics but of any wave mechanics in general.

When we apply these conditions to the orbitals of electrons around atoms an interesting result arises. The crests between waves on the same waveform are equidistant by mathematical definition. Thus in order for the waves to successfully wrap around an atom such that they match in a continuous fashion on the other side of the atom, the crests must be equidistantly separated from each other in three dimensions around the atom. If these crests are probability amplitudes though, then when they collapse they will most probably form particles at the peaks of the crests rather than anywhere else.

If these equidistantly spaced points are connected though, they form three-dimensional solids. A solid with equidistantly spaced vertices and identical polygonal faces is a Platonic solid. As the atomic orbitals obey these principles of wave mechanics, it follows that the shapes of electron orbitals should be directly connected with the geometry of Platonic solids also, and in reality this is what is found.[2]

The Platonic solids come up elsewhere in physics as well. It is interesting to note that the Platonic solid Plato associated with the ether, or what we would now call space-time, is the dodecahedron. The dodecahedron is the Platonic solid having twelve pentagonal faces formed by twenty vertices.

The dodecahedron also just so happens to have arisen in modern cosmology with the discovery that the shape of the universe may be a hyper-dodecahedron. This discovery was made by the French physicist Jean-Pierre Luminet after carefully observing the cosmic microwave background, the radiation left over from the Big Bang.[3] These observations showed that the fluctuations in the cosmic microwave background were different than one would expect if the radiation were travelling normally through infinite space. In an infinite space you would expect to see fluctuations of all sizes. However, Luminet found that this was not the case, and that there was a cut-off in wave-length. Furthermore, the structure of these fluctuations was such that it only made sense if the universe were a dodecahedron.[4]

It is interesting to note that the dodecahedron is directly associated with a geometry we have looked at before under a different context; the pentagon. In Chapter IV we saw that the pentagon inscribing a pentagram was directly associated with the fundamental structure of space-time in CDT through the two-dimensional projection of four-dimensional pentatopes. If this is the case, then it may be that the dodecahedral shape of the universe is the byproduct of space-time in the very early universe naturally forming into this shape by being built out of the pentagonal projections of these pentatopes.

Lastly, if the pentagon is directly built into the fundamental structure of space-time, then it provides a natural link built everywhere into reality to another pattern of sacred geometry; the golden ratio. The golden ratio shows up everywhere in nature at all scales, from the shape the human face, the growth of a conch shell, the arrangement of sunflower seeds to even the shape of galaxies. Thus it is natural to assume that it is built into nature somewhere at a fundamental level.

The golden ratio is built directly into the geometry of a pentagon. Each angle within a regular pentagon is 108 degrees. When this angular measurement is halved and its sine taken and then multiplied by two it replicates the golden ratio, denoted phi or φ:

$$2 \times \text{Sin}(108°/2) = 1.618034... = \varphi$$

Given that nature is built up at the most fundamental level from structures containing this ratio it is no surprise then that the golden ratio, which is itself scale-invariant, would appear at all scales in nature. Or as *The Emerald Tablet* would say:

"The structure of the microcosm is in accordance with the structure of the macrocosm."

By no means does this exhaust the topic of sacred geometry. In fact, its study is open-ended, and not all of it has been integrated into a proper scientific framework as of yet. For sake of space we must move on to examine how these geometries appear in living systems as well.

ii. Entanglement and Biology: Before discussing how it is possible for the quantum vibrations of Hermetic physics to manifest in biological systems it is important to discuss the presence of quantum phenomena in biology. Once quantum effects were thought to not be involved in biological systems at all due to biological systems being presumably too warm, wet and large for quantum mechanics to be relevant.

This was discovered to be simply wrong in recent years. Thus far quantum effects have been found to affect bird migration,[5] in the form of quantum computation in photosynthesis,[6] and in the sense of vision in phototransduction,[7] among other examples. The entire field of quantum biology, devoted to studying these sorts of quantum effects in biological systems, has come up in response to this. Thus it is entirely possible for quantum phenomena to affect biology in a manner that would give rise to the sort of occult geometry in the body that we are interested in exploring.

If occult geometry is intrinsic to biological systems then patterns of holistic quantum coherence should be central to biological systems. This holistic quantum coherence would exist as a large network of quantum entanglement linking many separate parts into a whole. If entanglement is central to biology though, then we should find it in the most important core of biological systems, their DNA.

In fact we do. In 2011, it was found that DNA is held together in the nuclei of cells by quantum entanglement.[8] This explains why DNA decays so rapidly outside of the nucleus and yet somehow manages to maintain its integrity within the nucleus. Given that DNA is replicated together prior to mitosis it is possible that this DNA is not only entangled within their individual nuclei but between nuclei in different cells as well.

A possible piece of evidence to suggest that DNA would be entangled between cells came in 2009 in the form of claims of DNA transduction or DNA teleportation by French Nobel laureate, Luc Montagnier.[9] This claim is controversial as the time scales on which it was supposed to have happened were not right for quantum phenomena.[10] Whether or not this is the case, the concept appears to be correct, even if the particular results Montagnier reported are not.

Perhaps a stronger piece of evidence though is the recent discovery of the real nature of DNA transcription in the nucleus. It was long thought that DNA is transcribed in an orderly fashion with an entire strand of DNA transcribed at once. Researchers recently caught DNA transcription on video and discovered that rather than a neat transcription of a single strand of DNA, the DNA molecule broke down into many strands.[11] Some of these strands transcribed faster or slower than others. Others started or stopped at certain points and then started up again. All of this resembled a very busy and complex mess, but when the process was done an exact copy of the DNA was produced again.

Given the high level of correlation of these dissociated fragments it seems likely they there were orchestrated through something like quantum entanglement. And given that there was nothing in the vicinity to explain why they were doing this, it is entirely plausible to suggest that this was due to entanglement links not even located in the cell's nucleus itself, raising the possibility that it was due to entanglement in other cells.

Due to the difficulty in testing for quantum effects in biological systems it may take some time for such a thesis to be confirmed. However, given the occult connections with quantum phenomena as seen in previous chapters on the one hand and the apparent link to occult patterns in the living body on the other, it is reasonable to think that these patterns exist even if they have not been found by modern science as of yet. We may only need to wait for future decades of research.

There is also another possibility. Previously we discussed how the collapse of the wave-function and thus quantum coherence and quantum entanglement are relative to the reference frame of the observer. It is often assumed that the environment in which the collapses and entanglements occur is objective, but this is an arbitrary distinction to make due to the relational nature of quantum phenomena. This being the case, there are two possibilities:

1. A pattern of entanglement exists in living organisms via hidden quantum biological structures that have only been partially discovered, *in an objective environment*.

2. A pattern of entanglement exists in living organisms *relative to the reference frame of the conscious observer*.

While it may be that further research reveals 1 to be correct as well, there is some basic sense in which 2 is already correct. Due to both realism and macrorealism being falsified, the body would naturally be in a giant macroscopic superposition from the subjective reference frame of the observer until he or she gains some external sensation of it. In fact, only the body and the external world would be "real" relative to the reference frame of the observer insofar as he or she has perception of it. The rest of it existing in a non-perceived state would be in superposition in a waveform. Given that the occult phenomena said to be associated with geometric structures in the body are of a phenomenological variety anyway, this seems like a natural assumption.

iii. Chakras, Acupuncture Points, and Meridians: If the body is treated as a coherent system of resonant standing waves then these waveforms should automatically produce points of constructive interference. These points would then be arranged across the body in certain patterns as determined by the geometry of the underlying waveforms.

Of course none of this would be visible to direct empirical detection via medical technology as that would defeat the entire purpose. The waveforms creating them would collapse, generating the image of a body or of organs and systems in the body. These points would exist as structures in the Hilbert space "shadow" of the body, rather than in physical space, and only from the reference frame of the observer.

Due the consciousness in the inner mental plane being fractured though, the observer would not be directly aware of them. However as we will see shortly, inner mental or emotional states are sometimes correlated with these points in the phenomenology of the observer.

One obvious example of these points comes to mind with the seven chakras aligned down the base of the spine. These chakras are arranged proportionally with respect to each other according to the golden ratio. The exact mathematical reasons for this are not understood, but likely relate to scale-invariant heterodyne waveforms existing across the human body, as these preserve the phi ratio.[12]

It is interesting to note that each of these chakras is associated with an emotional or spiritual state. For instance, the heart chakra is associated with love, and is "located" as a Hilbert space projection over the heart. When energy, in the form of quantum vibrations are flowing through it, the person is experiencing love.

From the phenomenology of the subject this is quite interesting. If we were to think in terms of modern materialistic science we would think that the feeling of love, as with all feelings, is something generated in the brain. Yet somehow when we are in love, and we are asked where we think the feeling of the sensation of love is coming from as a purely phenomenological matter, we will undoubtedly say it is coming from our heart. Likewise, when we have feelings of power we feel them coming from the solar plexus. Only this is where the solar plexus chakra, which is associated with our power, is located.

It is strange that these feelings should be associated with any organ of the body other than the brain, and if not the brain, then why not the lungs, the liver or the stomach? This matching of phenomenology to specific points in the body makes sense though if we treat them as constructive interference points of quantum waves with carrying actual cognitive states such as sensations of love or of power.

The sensation associated with the crown chakra is that of spiritual connection with God. This is at the top of the head, which would make sense if the body is to be entangled with the universal wave-function containing God's mental state, rather than being a separate quantum cognitive state whose waveform ends at the end of the body.

Similarly, these constructive interference points can explain acupuncture points. Meridians would be the lines showing the locations of smaller waveforms in the body connecting these points. The "qi" or "vital force" said to flow along them is the vibrational energy as perceived mentally in the phenomenology of the subject, which is similarly encoded in the waveform due to quantum cognition. To corroborate this thesis, chakras and acupuncture meridians have since been shown to have specific frequencies, and that if these are disrupted, it will adversely affect the patient, thereby revealing them to be phenomena based in wave mechanics.[13]

iv. Ley Lines and Energy Centers: Just as there can be points of constructive geometric interference mapped to the body as projected by the wave-forms in the mental and spiritual planes so too can there be similar patterns on the earth as a whole. Of course these points would not be visible within physical reality, but would rather exist as a mapping to various geographic points from within these planes. Additionally, the waveforms associated with them could only exist at the level of the universal wave-function. In any other reference frame the environment would be immediately collapsed by whatever else is not contained in the wave-function of the remaining system. However, as wave-functions decompose into smaller wave-functions, the effects of these points in the universal wave-function would still influence the region.

Just as acupuncture points and chakras focus energy in the "shadow projection" of the body in mental plane behind space-time, so too would there be energy centers on the earth produced in a similar fashion. Likewise, these would have "spiritual charges" associated with them, which would not be visible in a materialistic what-you-see-is-all-there-is view of the world. The notion of a cursed location or a haunted house makes a lot more sense in light of this, as does the notion of consecrating or blessing a site. Extending this concept further, the waveforms connecting these points or energy centers would correspond to what are referred to as ley lines.

With this knowledge in mind, it should be noted that many sites of spiritual or religious significance are located at these points. One of the largest such nodes is located in Giza where the pyramids are constructed. Another on this same grid is Stonehenge, and an investigation will reveal that many other significant sites are connected by the same means.

It is also of interest to the reader to note that this knowledge is already possessed by secretive organizations and is used in the construction of cities and buildings for purposes of harnessing these spiritual powers for nefarious ends. The activities of these individuals has gone on largely unnoticed, and they have amassed much power over humanity through occult means. This is mostly due to general ignorance among the public about the spiritual significance of their activities and about the nature of reality which allows them to manipulate such forces. Knowledge is power however, and given that the reader now knows the mechanisms by which power is taken, he or she can help take it back.

C. DNA And Enochian Magic

The Principle of Vibration has another occult application related to biology besides geometrically arranged energy points. This is related directly to the genetic code in DNA. The story behind this relates to a pair of Renaissance era occultists named John Dee and Edward Kelley, who appear to have mentored Shakespeare.[14]

Dee and Kelley claim to have communicated with angels in a language Kelley referred to as "angelic." This angelic language was later named Enochian. The angels attempted to teach them this language, and since then direct matches to genetic coding have been discovered with this language.

The Enochian alphabet has twenty-two letters. These correlate to the twenty amino acids coded for by DNA codons as well as the start codon and stop codons, instructing the RNA to either begin translating the genetic code into proteins or to stop.

The angels instructed Dee and Kelley how to produce a "seal of truth" from these letters in the form of an eight by eight grid of Enochian letters. This gives us the letters arranged into sixty-four squares on the grid. The parallel to genetics is apparent in that the coding of amino acids is done through sixty-four codons, with some amino acids being coded for by multiple codons.

These parallels are detailed enough that is has since been possible to construct a table of Enochian letters matching the characters in the "seal of truth" to their corresponding DNA codons. Additionally, this has been matched to the sixty-four hexagrams of the I Ching, as well as correspondences to the Mayan Tzolkin.[15]

If DNA is in a state of quantum coherence in the nucleus, then this should come as no surprise. Each codon would have a specific quantum wave-function with a specific rate of vibration based on its mass in the virtual physical plane. However due to the aforementioned occult properties of quantum wave-functions discussed in the preceding chapters, each wave-function should possess an internal mental state and with that a phonemic representation as well, such as was seen in the discussion on magickal languages at the end of Chapter IV.

These phonemic representations would then correspond to letters in a magickal alphabet in a magickal language such as are used for spells. In this case, the magickal language coding for DNA would be Enochian, and genetics could presumably be manipulated through the use of this sort of language. Again it should be stressed that the authors are not suggesting that this should be used, only that the reader be made aware of its existence.

On the topic of genetics in relation to Hermetic science it is also worth briefly mentioning the topic of DNA activation. This is an artifact within the "virtual reality" of space-time of the fractured consciousness in Hilbert space described in Chapter IV comprising the fallen state. This is what is commonly referred to as "junk DNA," which appears to have no biological function, but in fact codes for innate human abilities that have been suppressed due to the fallen state.

Sometimes this DNA is activated often in association with spiritual awakening. When this occurs the subject will experience a rubbery sensation in parts of his or her body that last for prolonged periods of time while the process occurs. The subject should not be alarmed if this occurs though as it is itself harmless and will subside in time. Due to the archontic presence on this planet, this phenomenon is relatively rare, as these archons seek to suppress humanity in the lower vibrations of this fragmented condition.

D. Gematria

The last topic we will cover in relation to vibration is gematria. Gematria is a subject closely associated with the Kabbalah, and with numerology. However like the preceding topics it has a very rational explanation once properly understood within the framework thus far provided.

The idea with gematria is that a person's name, or a named location, or so forth, has a numerical coding associated with it that can be used to correlate that person with other persons, places, life events, personality characteristics and so forth with a similar encoding. The idea is that the person is likely to be correlated to these persons, places, and events in real life. This appears on face value to be a purely mystical connection of a name with numbers, and in a certain sense it is. However unlike as is commonly thought, it can be understood quite logically in terms of the principles of Mentalism, Correspondence, and Vibration.

As discussed earlier, the phonemes, such as what comprise names, are the auditory representations of cognitive states which exist in Hilbert space. Because of this each exists as a quantum state with an associated vibration. If parents are not capricious in naming their children, but give their children names that they feel are significant to each particular child, then that feeling is likely to be guided by the collective unconscious. Thus a person's name will roughly match his or her actual vibration.

This is also the basis behind the occult belief that the name holds the power of what is named, as this is literally the case in a certain sense. The name holds the vibration of the named, by which the named is controlled.

This vibration can be indexed in terms of some number, similar to how library books are indexed. Gematria then acts as an index to correspond all of the persons, places, and things catalogued by their vibrations in this fashion with each other.

Within the physical world these persons, places, and things may appear to have no obvious connection to each other. However, in Hilbert space, which is what programs the "virtual world" of emergent space-time, they will have a close connection to each other. As the system's "program" is literally organized by these vibrations, there is some probability that it will correlate persons places and things with the similar "catalog numbers," and thereby produce connections between them within the virtual reality. As this "program" is not in physical reality, these connections will appear "mystical" when in fact they are merely correlated at a deeper level of reality.

Many other phenomena are associated with the Principle of Vibration, just as they are with Mentalism and Correspondence, due to the universal ubiquity of the phenomenon of vibration. However, we feel what we have written here is sufficient to let the reader get a general grasp of how vibration is applied to understand a variety of occult topics, and we do not have the space to go on. Instead we will now move on to the Principle of Polarity, which is related to the Principle of Vibration anyway.

VII The Principle of Polarity

The next principle to study after the Principle of Vibration is the Principle of Polarity. This principle relates what happens as vibration is increased or decreased along a scale. As we learned in Chapter V, each state of being has a vibration of its own, and as we learned in Chapter II, these states of being are mental. Thus this scale of vibration relates to mental states. In *The Kybalion's* statement of the Principle of Polarity the extremes of these scales are called poles, hence the name "*polarity:*"

"The great Fourth Hermetic Principle—the Principle of Polarity—embodies the truth that all manifested things have "two sides"; "two aspects"; "two poles"; a "pair of opposites," with manifold degrees between the two extremes."

The Principle of Polarity tells us that these scales of vibration relate to corresponding scales of mental states, such that all of the mental states on a scale are related to each other with poles existing at each extreme. All of the mental states on one scale are degrees of each other varying between the extremes of the poles, as one would expect given that it is nonsensical to speak of loudness coming after coldness, or hardness coming after the color yellow:

"When, however, the Principle of Polarity is once grasped, and it is seen that the mental changes are occasioned by a change of polarity—a sliding along the same scale—the matter is more readily understood. The change is not in the nature of a transmutation of one thing into another thing entirely different—but is merely a change of degree in the same things, a vastly important difference. For instance, borrowing an analogy from the Physical Plane, it is impossible to change Heat into Sharpness, Loudness, Highness, etc., but Heat may readily be transmuted into Cold, simply by lowering the vibrations. In the same way Hate and Love are mutually transmutable; so are Fear and Courage. But Fear cannot be transformed into Love, nor can Courage be transmuted into Hate. The mental states belong to innumerable classes, each class of which has its opposite poles, along which transmutation is possible."

The Kybalion also relates the Principle of Polarity to two seemingly contradictory halves of paradoxes being ultimately resolvable. This is true too as we will explore at the end of this short chapter. However, first we will examine the simple physics behind this principle.

A. The Physics of Polarity

The physics behind the principle of polarity can be directly understood from the physics of vibration and mentalism. Specifically, all that is needed is the basic understanding of the quantum waves described in Chapter V, and the physics correlating cognitive states to quantum states or to entanglement described in Chapter II.

In Chapter II we examined three theories all leading to the same general conclusion relating mental states to the coherent pre-collapse state of the wave-function. By further examining the physics of these wave-functions in relation to these theories, Polarity can be reproduced with all three of them in a trivial fashion. We will reexamine these three here towards that end.

1. Conscious Realism: On conscious realism the wave-function is derived by the interaction of conscious agents.[1] More conscious agents can be added to a wave-function, thereby entangling them into a larger conscious agent.[2] As entangled systems have a higher vibration than single systems, the vibration would increase as more conscious agents are added to the system. In turn these would constitute new larger conscious agents with new internal distinctive conscious states of their own. Thus as the vibration increases new conscious states appear.

2. Informational Idealism: Similarly, if we apply Toker's informational idealism to fundamental physics it would use quantum integrated information or quantum phi, which is measured in terms of entanglement.[3] As the entanglement is increased, new phi states with more information are created, with a subsequently greater degree of vibration. As was noted in Chapter II, it is uncertain whether the concept of phi can be extended into the quantum realm due to conflicting calculations of quantum phi.[4,5] Though this is unsettled, if quantum phi is a valid concept, the greater phi to greater vibration correlation would correspond to the production of new conscious states as well, replicating Polarity.

3. Quantum Cognition: The case of quantum cognition is generic and in principle encompasses both of the prior models. Though it also derives the same result independently. If each cognitive state is encoded in a wave-function as explained in Chapter II, then a wave-function containing only that cognitive state will have a vibration specific to that cognitive state. Changing that vibration will therefore necessarily correspond to a different cognitive state.

In each of these cases, the physics says nothing about the mental states of these wave-functions being of the same kind and on a scale between two poles. However, this is because the physics is being applied generically to cognitive states of any kind. Naturally though along the way more cognitive states of the same kind will appear along the scale of vibration, and these states will be closer to one pole or another.

It may be that the vibratory frequency encoding the concept of yellow in the mind is anteceded by the vibration of loudness or softness, as this is not yet determined. However even if it is, the vibration encoding the concept of another color, say blue or red, will definitely be on a different vibration. As this vibration will necessarily be either below yellow or above it, it would follow that a scale of these vibrations matches to a scale of color, just as polarity tells us. Of course color is just one example, but the reasons given here apply universally to any mental state if the preceding physics of quantum cognition is true, thereby reproducing the Principle of Polarity.

B. Quantum Smell

We would be negligent in our duties to provide the physics of Polarity without also giving a notable real world example already established by modern science. In recent years, quantum biology has given us a fascinating example of Polarity in the discovery of quantum smell. Here different smells, which are olfactory sensations of the *mind*, have actually been shown by science to be the direct causal result of different quantum vibrations.

Our olfactory senses are due to compounds with specific smells fitting or not fitting into chemical receptors in our noses. Thus whether or not a certain chemical fits into the receptor is determined by the shape of the molecule and the shape of the receptor. If the molecule is of the wrong shape it will not fit into the receptor, and thus a certain smell will not be produced. Likewise, a molecule fitting into a particular receptor will result in a certain smell being produced.

One would think this is all there is to smell, but this is wrong. As it turns out the same kind of molecule can be fitted into the exact same receptor and yet two different smells can be produced. This result was discovered by the Greek scientist Luca Turin in experiments with acetophenone and fruit flies.[6] Acetophenone is an ingredient commonly found in perfume, and one would think that two chemically identical acetophenone molecules would smell the same. This is not true though.

It is possible to change the quantum vibration of one of these molecules by making it more massive than another, and it is possible to do this *without* changing the chemical composition of the molecule. Protium and deuterium are both different isotopes of hydrogen, an element found in acetophenone. Protium contains a single proton in its nucleus, while deuterium contains both a proton and a neutron. Due to this, deuterium is heavier than protium.

When one plugs this difference into the Schrodinger equation, each of these produces wave-functions with different vibration rates. When these different isotopes are used in acetophenone therefore, the wave-functions of acetophenone molecules likewise have different vibrational rates.

The fruit flies in Turin's experiment were actually able to smell this vibrational difference when these different acetophenone molecules with different vibrational rates interacted directly with their central nervous systems. This result has been popularly known as "quantum smell" in the press. Given that the only factor involved in altering the smell from one smell to another was the molecule's quantum vibration, this presents a direct scientific confirmation of the Principle of Polarity in modern science.

C. Mental Induction

The Kybalion mentions the topic of mental induction also as being related to Polarity. This principle shows how it is possible for one person to influence the mental state of another through vibration. This is found in its statement here:

"In addition to the changing of the poles of one's own mental states by the operation of the art of Polarization, the phenomena of Mental Influence, in its manifold phases, shows us that the principle may be extended so as to embrace the phenomena of the influence of one mind over that of another, of which so much has been written and taught of late years. When it is understood that Mental Induction is possible, that is that mental states may be produced by "induction" from others, then we can readily see how a certain rate of vibration, or polarization of a certain mental state, may be communicated to another person, and his polarity in that class of mental states thus changed."

This is perhaps very intuitive even for those who are not aware of Hermeticism, and can be readily seen in the effects of politicians and orators. The orator changes the tone of his speech as well as his mental state to convey a certain emotion. So long as the emotional state of the audience is already resonating with his speech, they can be swung emotionally in his favor. Likewise presenters can produce similar effects in their audiences based on subtle appeals to emotion throughout.

Of course this is obvious and goes without saying. However, it is worth noting that this works on a deeper collective unconscious level as well. Enough people being swayed in a certain way will produce an unconscious resonance between them, the tone of which can be picked up in others. These others then will be more emotionally susceptible to this resonance simply because of its presence in the collective unconscious.

D. Polarity and Paradoxes

Lastly there is the issue of Polarity being related to the resolution of paradoxes in *The Kybalion*, which we mentioned at the beginning of the chapter. It is worth quoting what *The Kybalion* has to say about it here:

"Man has always recognized something akin to this Principle, and has endeavored to express it by such sayings, maxims and aphorisms as the following: "Everything is and isn't, at the same time"; "all truths are but half-truths"; "every truth is half-false"; "there are two sides to everything"; "there is a reverse side to every shield," etc., etc.

The Hermetic Teachings are to the effect that the difference between things seemingly diametrically opposed to each other is merely a matter of degree. It teaches that "the pairs of opposites may be reconciled," and that "thesis and anti-thesis are identical in nature, but different in degree"; and that the "universal reconciliation of opposites" is effected by a recognition of this Principle of Polarity."

This follows naturally from Polarity if we recognize that a paradox, and likewise its resolution, are among other things mental states. For example, even earlier in this book the reader was introduced to seemingly different and conflicting concepts in religion in relation to both the afterlife and the fallen state. Prior to reading these sections, the reader may have had a *mental state* wherein he or she thought these were mutually exclusive and incompatible ideas. Naturally this *mental state* would have had a vibration.

Likewise upon learning new knowledge, the reader would have had another *different mental state* wherein he or she was aware of how these apparently differing descriptions were in fact analogous to blind men disagreeing over the correct description of an elephant. This different mental state would have had another different vibration as well. Thus as the vibration was changed, in this case increased by an increase in understanding, the paradox became resolved.

The ultimate point of bringing this up is that there are no true paradoxes. Excluding what are deliberate lies, no actual contradictions can exist within The Mind of The All. God is Truth manifest, and is also One. This Truth underlies logic and can not contradict itself. Thus in reality there are no true paradoxes, only the one Truth seen in part by finite minds from finite perspectives.

Polarity is an interesting topic in its own right with other applications as well. Though now we must move on to another principle, the Principle of Rhythm. As the reader might expect, this principle also arises as a direct consequence of the Principle of Vibration.

VII The Principle of Rhythm

The Principle of Rhythm is one of the more commonsensical principles of Hermeticism and can be derived directly from Vibration. As stated in *The Kybalion*, it simply says that:

"Everything flows out and in; everything has its tides; all things rise and fall; the pendulum-swing manifests in everything; the measure of the swing to the right, is the measure of the swing to the left; rhythm compensates."

The reader is probably already well familiar with this as day to day life carries with it many such rhythms and repeating cycles. However, this is also built in at the fundamental level of reality if the obvious consequences of the Principle of Vibration, which is really only the science of quantum mechanics, are examined. As we will see along the way it is also connected to the Principle of Polarity.

A. The Physics of Rhythm
Rhythm is also referred to in scientific terms as frequency, the rate at which a number of waves moves over a period of time. The connection to quantum mechanics is obvious here as wave-functions are waveforms, which by definition have their own vibratory *frequency* or *rhythm*. Thus Rhythm is built at its most fundamental level directly into Vibration.

To fully understand the Principle of Rhythm, we need to see how this also scales up from fundamental vibrations. Wave-functions are described in terms of sines and cosines. Any combination of sines and cosines in turn, comprises a larger waveform that is described as a Fourier series. Since everything, including all particles and even space-time itself, is described by wave-functions, these waveforms would repeat on all scales in everything as Fourier series.

Waveforms have crests and troughs where they experience minima and maxima. This would of course correspond to the tides and rises and falls described by Hermeticism that everything has. A waveform has a single frequency however, and thus it should be curious that *The Kybalion* should describe it as connected with Polarity, which deals with the decrease and increase of frequency along a scale:

"Rhythm manifests between the two poles established by the Principle of Polarity. This does not mean, however, that the pendulum of Rhythm swings to the extreme poles, for this rarely happens; in fact, it is difficult to establish the extreme polar opposites in the majority of cases. But the swing is ever "toward" first one pole and then the other."

At first glance this may be a bit perplexing, but with some thought it becomes obvious how the two are connected. In a Fourier series there are many waveforms superimposed over one another. Some of these waves are large and others are smaller. The combined average frequency of the composite waveform they create is based on their collective energy. As there are many waves of many frequencies comprising this waveform, this averaged frequency may not always be evident at all points in the waves. For example, a trough of a very large wave could appear, such that at a sufficiently limited scale, the contribution to the waveform of that particular wave is negligible. Here the remainder of the waveforms in the pattern would become more evident, and the contribution of the larger waves energy would not be as apparent. Thus for this shorter span of the waveform, the averaged vibration of the waveform would be more diminished than that of the waveform on a whole. Since Polarity is described in terms of changes in vibration, polarity would naturally rise and fall with the crests and troughs of a complex waveform.

This would in turn correspond to a rise and fall in the mental states corresponding to such vibrations. The probability of mental states occurring would still be there though. However, since the probability in quantum mechanics of a state occurring is dependent on the amplitude of the wave-function at any particular point, its chance of occurring would be diminished during such troughs.

These wave-functions as they exist as part of the environment are very tiny once the wave-function has been collapsed. Prior to that though the environment would exist as such a giant waveform with its own rhythm, which could in principle, as we will now see, last great spans of time. Despite being collapsed from the frame of the observer's environment, many resulting patterns, and conscious or subconscious influences would continue to exist and be related to the rhythms of the environment's wave-function from which they were produced or decomposed.

B. The Cycles of Ages

A significant occult phenomenon associated with Rhythm is the Cycle of Ages. This is the idea that the world travels through ages of time as if on a grand clock. The beginnings and endings of these ages correspond to the hidden but natural occult influences produced by these natural rhythms, which due to their mental nature and broad influence, elude the materialist and the reductionist.

Nevertheless, cycles of this nature do exist and have been observed and documented by many cultures. In the west they are the astrological ages, in the east and India they are the yugas, and the Mayans based their Tzolkin calendar on them. Not all of these refer to the same cycles, but they all belong to the same concept and it is possible to find links between them. Sometimes these cycles are associated with the zodiac. Though this does not have any additional significance beyond the fact that observations of the heavens are useful in establishing the existence of yet more corresponding cycles of rises and falls.

The correlation of the spiritual influence on these eras is no accident, and can be very clearly seen. The Age of the Taurus the bull for example ended with Moses destroying the golden calf and ushering in The Age of Aries the ram.[1] The association of the symbol of the ram's horn as a spiritual representation of the Law of Moses will be recognized by most readers. Likewise the Age of Pisces the fish, corresponds well to the Christian era, which is closely associated with the symbol of the fish.

The last era of the Mayan long count calendar, which was just recently exited earlier this decade, illustrates yet another notable example of the cycles of ages at work. The Mayan long count calendar is divided into four-hundred year eras or *baktuns*, and each of these eras has an associated theme. The theme of the previous era which lasted over the previous four centuries from 1618 AD to 2012 AD was that of the "triumph of materialism" or the "transformation of matter."[2]

This was known about prior to any of the intellectual developments going on in Europe at the time, which since the time of the Renaissance and The Enlightenment have been increasingly materialistic in their thinking. In this case, the spiritual underpinning behind this development was clearly based in and predicted by the cycles of ages recorded in the long count calendar.

This period of time and of the prevalence of materialistic thinking and its consequences was predicted by other such cycles as well, sometimes with uncanny accuracy. Our materialistic era for instance also strongly correlates with the Kali Yuga of India.[3] Neither of the authors consider this to be a good development. Though this period has also produced many scientific and technological advancements which have also benefited humanity, the underlying spiritual and intellectual framework in which it has done so has left humanity severely impoverished and nihilistic in many respects. These times correlate to an era of forgetting and ignorance of the true nature of reality.

These two things need not be associated though. As has been previously described in this short book, a scientific view of the world need not also be a materialistic view of the world, despite the common but false conflation between the two. It should be noted that Hermes Trismegistus even prophesied of this same era in the *Divine Pymander*.[4] The reader should also understand that the rise of materialistic thinking is not merely an inevitable result of progress, but of the spiritual influence of our age which will subside in time.

We will return to the topic of the age of materialism in the final chapter of this book as it is of great importance for the future. For now, we will turn our attention to another subject, that of the Principle of Cause & Effect.

VII The Principle of Cause and Effect

Causality is intuitively obvious to most. Thus, it is curious that such a readily available intuition should be considered hidden enough such that it appears in *The Kybalion* as one of the Hermetic principles. Nevertheless, it does:

"The great Sixth Hermetic Principle–the Principle of Cause and Effect–embodies the truth that Law pervades the Universe; that nothing happens by Chance; that Chance is merely a term indicating cause existing but not recognized or perceived; that phenomena is continuous, without break or exception."

The reason for this is that while on the surface the Hermetic Principle of Cause and Effect does include the common concept of causation, it is more primarily concerned with cause and effect on a much deeper level. This can be seen elsewhere in *The Kybalion's* writings on Cause and Effect:

"there are many planes of causation, but nothing escapes the Law."

This indicates that the causality Hermeticism is concerned with is not primarily so much the mundane causes and effects of ordinary mechanical laws, but rather that of causes and effects occurring between levels of reality. Because of this it is directly correlated to the Principle of Correspondence, which describes those levels of reality. As a consequence, the explanation of Cause and Effect follows as an extension of the explanation of Correspondence.

A. The Physics of Cause and Effect
In Chapter III the reader saw how the Principle of Correspondence could be accounted for in terms of physics. The world we see is not material, but is actually a simulation produced from a deeper level of reality, a quantum probability "space," which physicists call Hilbert space. The reader then saw that when quantum cognition is combined with this understanding, that this deeper level of reality possesses a mental nature in which our own minds take part.
 It was further seen that most of this deeper level of reality is off limits to our waking minds, but exists in our subconscious. Thus it is possible to equate it with Jung's collective unconscious, and its contents with Jungian archetypes and forms. As these archetypes were the mental programming behind the physical plane, it follows that so too they would form correspondences with the contents of the physical plane. This was related to the phenomenon of synchronicity which we will examine later in this chapter. The Principle of Cause and Effect simply takes this one step further. If the contents and occurrences of the physical world correspond to the archetypes and occurrences in the mental world, then so too changes in the mental world should correspond to changes in the physical world. Given that the mental and spiritual planes are more fundamental than the physical world, it follows that these changes should be causative as well. A cause in the mental or spiritual planes should correspond to an effect in the physical planes.
 Likewise it is conceivable for actions on the physical plane to shift the programming behind the simulation so as to affect the goings on within the mental and spiritual planes. This does not mean that cause and effect is going in reverse. Rather it only means that that mental and spiritual correspondents of physical actions are themselves influencing secondary mental and spiritual goings-on to produce

some corresponding effect in the physical plane. In this manner the explanation of Cause and Effect can be accounted for by the same science that explains correspondence.

One can easily see how this could account for such phenomena as telekinesis. If one's mental state is a quantum state as quantum cognition would suggest, it could in principle be entangled with the wave-function behind some external object. If one were then to focus mental intent on the "programming" behind that object, one could in principle move it. This is no different from how objects can move probabilistically in quantum tunneling, only that due to the mental nature of the wave-function, the tunneling would be directed through mental intention.

This sort of telekinetic phenomena would be a good test of Cause and Effect as it would operate on the same physics. Exactly such tests have been done by the Global Consciousness Project, or GCP.[1]

In these tests random number generators are distributed in various locations across the world and produce constant streams of random numbers day after day for years. These random number generators are based on the radioactive decay of atoms, which is a quantum probabilistic process and is thus truly random. However, during significant events which draw the focus of public consciousness the numbers recorded from the random number generators spike beyond what would be expected if there was not a causal correlation. During the events of September 11, 2001 for instance, one such spike was recorded with odds of one in thirty-five.[2]

As any one of these events is hard to establish proof of telekinesis by itself, the GCP has run trials continuously for over the length of its project. The cumulative result of doing this has achieved results of 7.31 sigma.[3] It should be noted again that it takes 5 sigma for something to be scientifically proven, and that the odds of something occurring at 7 sigma by pure chance is one in a trillion.

This means that the telekinetic phenomena observed by the GCP is more certain than the existence of the Higgs Boson.[4] If the reader believes in the discovery of the Higgs boson by the Large Hadron Collider based on the scientific evidence for it, then so too should he or she believe in the existence of telekinetic phenomena for the same reason. This would rule out any reason for not believing it as being unjustified metaphysical bias or a non-rational preference not to believe it. It would also mean that the odds that the telekinetic effect measured by the GCP is *not* occurring is *greater than* one in a trillion.

The direct causative link between internal mental states and the wave-function described here as the mechanism behind Cause and Effect has also been confirmed. A set of six experiments published in *Physics Essays* demonstrated just such a connection. These experiments were conducted by Dean Radin of the IONS and showed that the spectral pattern on a double slit experiment could be directly affected through mental attention given to it.[5] As this pattern is directly dependent on the wave-function, it demonstrates both the non-locality of consciousness within the wave-function, and the mind's causative effect on quantum probability.

Thus the Principle of Cause and Effect can be considered to be scientifically grounded if one rejects irrational biases. Now of course when one discusses the Principle of Cause and Effect there are the issues of free-will and quantum probability, both of which are said to flout causation. It is true that there is a sense in which they do, but not everything is as it seems. As we will now see, it is more accurate to say that they *bypass* Cause and Effect rather than violate it.

B. Free-Will and Quantum Probability

In any discussion of causation, it is necessary to examine the cases that are said to violate it. These are two-fold, free-will and quantum probability. The first is the freedom of the will of the person to act in a manner undetermined by prior causes. The second is a phenomenon of quantum particles to act in a manner that is in principle uncertain and undetermined beyond a certain range of values.

When we explore this topic from the context of the preceding Hermetic physics we find that these two exceptions are in reality one and the same. The free-will of a mind is the indeterminacy of quantum mechanics seen from a first person point of view. Once this is understood, it can be understood why neither actually violates Cause and Effect but rather only bypass it.

Firstly, we need to ask what this mind is that possesses free-will. Beyond the first person perspective we have in Chapter II been able to identify it as what comprises the quantum wave-function. This follows directly from what we learned regarding Hoffman's derivation of the wave-function from systems of interacting conscious agents, as well as indirectly from the other approaches previously described.

If this is the case, then the probabilistic outputs generated by these wave-functions would be identical with the actions of these minds. The quantum probabilistic states pre-collapse would in turn be the inner states of these minds in determining those actions. Thus within the framework of the physics we have discussed, the indeterminacy in free-will would be identical with indeterminacy in quantum probability.

How is this reconcilable with Cause and Effect? The answer to this follows directly from Mentalism. Causal chains always start with first causes. The first cause is necessarily the first thing that exists, The Mind of The All. The Mind of The All would not have this ability to be a first cause, unless this ability were innate in mind to begin with. Thus mind has the ability to act as a first cause, which itself is uncaused.

This does not violate Cause and Effect, as Cause and Effect says nothing about first causes. Rather it only applies to chains of causes and effects. Thus free-will and quantum indeterminacy do not violate Cause and Effect, but only creates the causes upon which Cause and Effect acts.

Another comment should be made at this point in regards to free-will being identical with quantum probability. The range of probability of a wave-function is dependent upon its energy value. For example, in the case of quantum tunneling, a higher energy particle will be more likely to tunnel across a barrier than a lower energy particle. But in Chapter V we saw that vibration is dependent upon energy. The higher the energy the higher the vibration.

Thus consciousness at a higher vibration has more free-will than consciousness at a lower vibration. This appears to be implied in the Kybalion as well, if we recognize that the "higher planes" are also of higher vibration:

"The majority of people are carried along like the falling stone, obedient to environment, outside influences and internal moods, desires, etc., not to speak of the desires and wills of others stronger than themselves, heredity, environment, and suggestion, carrying them along without resistance on their part, or the exercise of the Will. Moved like the pawns on the checkerboard of life, they play their parts and are laid aside after the game is over. But the Masters, knowing the rules of the game, rise above the plane of material life, and placing themselves in touch with the higher powers of their nature, dominate their own moods, characters, qualities, and polarity, as well as the environment surrounding them and thus become Movers in the game, instead of Pawns—Causes instead of Effects. The Masters do not escape the Causation of the higher planes, but fall in with the higher laws, and thus master circumstances on the lower plane. They thus form a conscious part of the Law, instead of being mere blind instruments. While they Serve on the Higher Planes, they Rule on the Material Plane."

This is also how the fallen condition described in Chapter IV is maintained. Such a state naturally limits the range of the free-will of those within it, and makes them susceptible to the influences of other beings who benefit from perpetuating their condition.

Lastly when discussing free-will, comments should be made concerning the experiments from neuroscience like those of Benjamin Libet that purportedly contradict free-will.[6] Experiments of this kind show subconscious neurological influences in the brain apparently preceding the conscious volition to act, thereby suggesting that free-will is an illusion.

This would all be true if we were to adopt naively materialistic views about the world. However as was discussed in Chapter II, the particles comprising the material world are not real before they are measured.[7] Furthermore, we saw how violations of the Leggett-Garg inequality scaled this non-reality up to the scale of macroscopic objects.[8] Thirdly, the non-reality of what is observed upon collapse is relative to and dependent upon the reference frame of the observer. Meaning if the observer does not see his or her own brain, then from *that observers reference frame*, the brain and its processes do not exist as real things as part of our physical reality, but exist in limbo as superpositions of quantum states.

When delayed choice quantum erasers are brought into the picture, this even applies to observations of past events.[9] Delayed choice quantum erasers are variations on the double slit experiment described in Chapter II wherein the decision on whether or not to measure the system is delayed until after the waves have gone through the slits. In these experiments it was discovered that the choice to measure them as particles after they had gone through the slits determined their past history as having been particles. This was even the case if they had gone through as waves before!

Because these antecedent brain processes are part of the same illusory physical plane it would follow that it is also governed by this physics. As a result, from the reference frame of the mind that makes these choices, the past histories of these antecedent brain processes would be just as illusory as the past histories in the quantum erasers. Thus experiments of this variety are incapable of contradicting free-will.

If this is the case though, then one may be tempted to ask whether or not other macroscopic objects in the physical plane also have fake histories, and why we have not seen evidence of this before. The fact of the matter is that we do, but we do not recognize it as such. This is none other than the phenomenon of synchronicity, which is directly related to Cause and Effect, which we will now turn to.

C. Synchronicity

We are all familiar with synchronicity, perhaps even on a day to day basis. Some events will occur of a coincidental nature, which are psychologically meaningful. We will think of someone and then get a call from them and so on. Sometimes these coincidences can be quite profound and defy the odds of probability. These are referred to as synchronicities.

These events appear to defy rational explanation, but in fact their existence can be explained very rationally if one recognizes the mental nature of reality. If the concept of quantum erasers just described is scaled up to the macroscopic level, then it follows that many things we assume to have reality are in fact in superposition with many possible histories behind them.

For example, before we open a door many possible people may be behind it. All of them are in superposition from our frame if we treat macrorealism as false. Each will have had a history of entering the room behind the door. If there is something guiding which state is selected from this superposition, and if that something is meaningfully related to our inner psychological states, then when we look, the person behind the door may "coincidentally" be someone who is meaningful to us. This mutual meeting would then seem surprising in a way that defies probability, and we would call it a "synchronicity."

All that is needed then is a mechanism to guide these probabilities in the superposition in a fashion that is psychologically significant to us. But as was just discussed in the previous section, we already do. The wave-functions containing these superpositions are minds. What superposition is selected is dependent upon the free and undetermined volition of these minds, which from an outside perspective, we call quantum probability.

As described in Chapter III, the entire collective unconscious can be explained in terms of these wave-functions. Thus conscious influences beneath our level of consciousness may directly select one outcome from among many. If these conscious influences are in our subconscious they would even be psychologically meaningful to us as synchronicities are described to be. It is also possible that if our consciousness is psychologically attached to *or entangled with* the situation strongly enough, that it may select these outcomes as well, creating synchronicities in a similar fashion.

Most people do not realize that their own consciousness has the capacity to do this. These occulted facts are not unknown to everyone though. In fact as we will now discover, they can be used to great effect to manipulate the consciousness of others and achieve desired outcomes as a result.

D. Sigil Magick

Sigil magick is closely connected with synchronicity, and it is far more ubiquitous than the average person realizes. It can be understood to work by the same mechanism used to explain synchronicity in the previous section. It simply puts to work in a controlled fashion what was otherwise the natural phenomenon of synchronicity.

The idea is that some symbol, referred to as a sigil, corresponds to something that actually exists in the subconscious. By putting the sigil in place, it can influence forces in the collective unconscious to influence the minds of other people or events by way of synchronicity.

On the surface level, these will appear to just be mundane symbols in the physical plane. In reality, when they are observed by people, they enter not only their inner mental states, but also by extension the larger inner mental states of the collective unconscious in which these smaller minds are contained. These larger inner mental states or archetypes are activated by these sigils, and can react by influencing the quantum probability space controlling the physical plane. A few examples of this are given here to illustrate how this works.

a. Sigil Magick In The Music Industry: So-called "lesser magick" which only influences the individual subconscious of the consumer to purchase a product is quite well known about. The reader has no doubt heard about this as subliminal advertising. However, as many vigilant observers have noticed, the music industry is engaged in the darker practice of "greater magick," the magick manipulating the world behind the physical plane. This involves the intentional placement of sigils and other occult symbolism in music videos for purposes of sigil magick.

What will occur is that sigils and other archetypal symbolism is placed into music videos. These symbols will then enter the consciousness of those watching the videos. As far as a perspective that is ignorant of these principles is concerned that is where it stops. Such a perspective is of course wrong.

What occurs next is that these sigils activate forces in the collective unconscious. Some of these forces are not merely impersonal either but can include various dark entities. These influences will then work upon the consciousness of those viewing these music videos unaware, effectively placing them under a spell or even under the influence of these dark entities in the collective unconscious.

Due to the negative nature of these influences, it is important for the public to both know the mechanisms described here by which they operate, and those who put them to use. Artists who engage in these and related practices are widespread, but include names such as Lady Gaga,[10] Beyonce,[11] Jay-Z,[12] Nicki Minaj,[13] and Kanye West[14] among many others. It should be noted though that most of these artists are only puppets, and that the real controllers are those behind the music industry.[15]

b. Summoning Kek With Meme Magick: The second example is interesting in that it involves sigil magick inducing actual synchronistic effects. These relate to the internet meme "Pepe the Frog," and its correlation to the alt-right movement and the seemingly im*probable* candidacy and election of Donald Trump.

Pepe the Frog became used as a symbol for alt-right activism on boards such as 4chan in the run-up to the 2016 US presidential election. There Pepe was associated with the internet slang term "kek." It was then discovered that Kek is actually the name of an Egyptian frog-headed deity. After this discovery, a number of highly unusual synchronicities were discovered in connection with this phenomenon.

A WordPress document entitled *The Truth About Pepe the Frog and The Cult of Kek*,[16] documents a variety of these. This includes frequent numerical synchronicities associated with 4chan discussion threads regarding Donald Trump, Pepe the Frog, or the internet slang term "KEK." Another example includes Egyptian hieroglyphs on an ancient figurine of Kek, displaying what appear to be someone on a computer entering memes.

Perhaps one of the most incredible of such synchronicities was the discovery of a record from the 1980's displaying a frog holding a magic wand.[17] The title of the record was P.E.P.E, the name of the aforementioned frog. Curiously P.E.P.E. stands for "Point Emerging Probably Entering." To make things weirder, the cover art for the album gave the appearance of the clocks on Trump Tower. If Kek is indeed behind this, perhaps this is the now-infamous frog-deity winking at us by alluding to the *probabilities* being manipulated in *probability space* to cause these effects to *emerge* into our *point* in the space-time realm as synchronicities.

Given the archetypal significance of a frog-headed being called Kek, it would appear that this is a textbook example of sigil magick in operation. The memes embed in people's consciousness and thereby activate Kek in the collective unconscious. The Kek being or archetype in the collective unconscious then influences the quantum probabilities programming the physical plane to produce these synchronicities. Whatever the case, once the magickal nature of this phenomenon was realized, it has since then only encouraged the spread of Pepe memes among enthusiastic supporters on internet boards.

Sigil magick is one of the means by which occult principles are used at the expense of the masses. Due to the fact that most do not know about these principles, it enables those who do to exploit this knowledge to their advantage. This creates a power differential between themselves and the masses. However, it is not the only means by which occult principles are used at the expense of the average man. We will discover another such abuse of the natural Hermetic principles occurring today in the next chapter on the Principle of Gender.

VIII The Principle of Gender

The Principle of Gender is the last of the seven Hermetic principles. It tells us that there are male and female principles present in everything. Here the term "principle" might perhaps better be described as an "essence" or "aspect." *The Kybalion* states it as follows:

"The great Seventh Hermetic Principle–the Principle of Gender–embodies the truth that there is Gender manifested in everything–that the Masculine and Feminine principles are ever present and active in all phases of phenomena, on each and every plane of life."

How this makes sense is difficult to understand at first, as our conception of gender is that it is a psychological or at least a biological phenomenon. This comes from the naïve surface level understanding of reality that comes to us from our phenomenal senses of physical plane alone. As *The Kybalion* explains, biological sex is only one particular manifestation of gender:

"A moment's consideration will show you that the word has a much broader and more general meaning than the term "Sex," the latter referring to the physical distinctions between male and female living things. Sex is merely a manifestation of Gender on a certain plane of the Great Physical Plane–the plane of organic life."

So what is this gender, and how is such a principle derived? Once we scratch the surface, we find that there is more to it than what we find in biology or psychology.

A. The Explanation Of Gender
The Principle of Gender has no obvious connection to physics or biology. This is due to the inbuilt limitations of science. Science operates on grounds of empirical evidence alone, and thereby by its very nature limits its scope only to what can be detected through the outer senses.
  In previous chapters, quantum cognition was discussed for example. Were it not for the correlation of quantum probabilities with the *third person reports* of inner subjective states, science would never have thought to match inner subjectivity with the mathematics of the wave-function, much less the possibility that this was anything more than a mathematical modeling. As the reader saw though, when the existing science was put into the light of a metaphysical derivation of Mentalism and Correspondence the connection between the two was obvious. The Principle of Gender is like this.
  i. The Principle of Mental Gender: In similar fashion, the understanding of the Principle of Gender has to do with the mental goings on behind the physical plane that are not directly accessible through empirical methods. To understand the origins of these goings-on as they relate to gender, it is best to examine what *The Kybalion* has to say about the related Principle of Mental Gender:

"These aspects of mind–the Masculine and Feminine Principles–the "I" and the "Me"– considered in connection with the well-known mental and psychic phenomena, give the master-key to these dimly known regions of mental operation and manifestation."

According to mental gender these two aspects of the psyche, the "I" and the "Me" are said to be masculine and feminine respectively. Why this is the case will be understood shortly. Firstly though, it is necessary to understand what exactly the "I" and the "Me" are, so that we can understand how they are related to one another:

"Each student is led to see that his consciousness gives him first a report of the existence of his Self–the report is "I Am." This at first seems to be the final words from the consciousness, but a little further examination discloses the fact that this "I Am" may be separated or split into two distinct parts, or aspects, which while working in unison and in conjunction, yet, nevertheless, may be separated in consciousness.

While at first there seems to be only an "I" existing, a more careful and closer examination reveals the fact that there exists an "I" and a "Me." These mental twins differ in their characteristics and nature, and an examination of their nature and the phenomena arising from the same will throw much light upon many of the problems of mental influence.

Let us begin with a consideration of the "Me," which is usually mistaken for the "I" by the student, until he presses the inquiry a little further back into the recesses of consciousness. A man thinks of his Self (in its aspect of "Me") as being composed of certain feelings, tastes, likes, dislikes, habits, peculiar ties, characteristics, etc., all of which go to make up his personality, or the "Self" known to himself and others."

The reader can see from this that the "I" refers to the irreducible self, divorced from any other content. The "Me" by contrast, refers to the various characteristics the "I" possesses, the memories, personality traits, and so on. When this is understood, the relationship between them is also understood.

The "I" possesses the "Me." When we have memories, emotions, personality traits and so on, we say that they are "within our mind." The mind of course is referring to the distinct irreducible self. Thus for purposes of understanding, we can think of the "Me" as being a circle contained within a larger circle which we will denote the "I."

However the *Emerald Tablet* tells us that the macrocosm is the microcosm writ large. Due to the Principle of Mentalism, this "I/Me" distinction applies to The Mind of The All as well, and everything contained within. With the context of The All in mind, we can then ask how the contents of The Mind of The All were created or were *engendered* in the first place.

The Mind of The All was the first thing to exist, and then it created from within itself. This produced the mental simulation that our physics has now discovered as described in the second chapter. These contents of The Mind of The All would be The Mind of The All's "Me." The cause for this creation would be the initiation of the "I" of The Mind of The All, and would be the product of the expression of the love of The Mind of The All, as all initiation is ultimately born out of love or desire. Likewise, the "Me" due to its very nature would be inherently receptive. These initiative and receptive principles are in turn the Masculine and Feminine Principles described in *The Kybalion*:

"The principle of Mental Gender gives the truth underlying the whole field of the phenomena of mental influence, etc. The tendency of the Feminine Principle is always in the direction of receiving impressions, while the tendency of the Masculine Principle is always in. the direction of giving out, or expressing."

ii. The Derivation of Gender: As all of reality is mental due to the Principle of Mentalism, any individual thing that exists would have to have been created in the same manner, and would thus be a manifestation of this pattern of gender. And in fact if we look closely, this sort of pattern is seen in nature. A good example of this is when we compare psychological gender with biological sex. These two things share commonalities that could only be explained as coincidental were the Hermetic Principle of Gender not understood.

In regards to *psychological* gender, a study[1] conducted in 2011 demonstrated that 83% of men wanted to initiate a relationship by asking someone out, with only 16% wanting to be asked out themselves. Similarly, 93% of women took on more receptive attitudes in relationships, wanting to be asked out instead, with only 6% of them wanting to ask others out.

This initiative/receptive pattern may be clear in psychology, but it is very evident in biology as well. If we compare the psychological patterns regarding gender to the biological facts regarding reproduction, we find a similar initiative/receptive pattern at work along male/female lines. To euphemize the facts regarding reproduction, the male plug goes into, signifying initiation, the female outlet that receives it. Likewise, the male sperm does the initiation in the creation of a new life, swimming to the female egg which similarly receives it.

These parallels make no sense in a materialistic paradigm other than as rather peculiar coincidences. However, they do make sense within the context of the Hermetic principles. The contents of the physical plane, in addition to the superficial material causes they are given, are also symbols of the deeper reality generating them as a mental construction. Underlying the physical plane are the mental and spiritual planes. Thus the goings-on in the physical plane are guided by the mental and spiritual planes, and are as such reflective of them as well, as we see in these parallels between psychological gender and biological sex.

Going further like this we see that biological sex is really only one manifestation of psychological gender. The distinction between biological sex as separate from the phenomenon of gender is a modern error due to an artificial materialistic understanding of the world.

iii. Equivalent Derivation of Gender with Physics: This principle could in principle be converted into the same physics used to describe Mentalism and Correspondence also, though such an explanation would hardly be necessary at this point. One would simply take the mentalistic metaphysical explanation given here and convert it into equivalent mentalistic physics.

This would be done by first identifying The Mind of The All with the universal wave-function. Then individuation of gendered units of consciousness would be explained in terms of smaller systems individuating from the universal wave-function in the form of disentangling with it. The universal wave-function would still exist in its own frame of course. However, the smaller minds below it would exist on lower "excitation levels" corresponding to lower vibrational planes or "emanations."

Given that the cognitive state of the universal wave-function would be in a unified state at the top level, any subsequent individuated systems would naturally complement each other as well. This would lead to these parts of the larger system being able to match each other along the lines of a gendered ordering of initiative and receptive attributes. The same initiative and receptive states unified together in the universal wave-function, or in any higher excitation states, would still exist and match each other on lower levels where they are individuated as distinct units. Thus this metaphysical derivation of Gender can be translated into a derivation based on equivalent physics.

The Kybalion uses the example of pairs of electrons and protons with positive and negative charge as examples of initiative and receptive principles in physical systems. Of course from a materialistic perspective these actions are seen as due to electrical forces, and this interpretation is not wrong. However, at a deeper level, the wave-functions forming the fields of these electrical forces are conscious agents according to Hoffman's description of the probability wave. Thus the actions of these fields can be seen as intentional as well, indicating at the most rudimentary level, an attraction or "love" between initiative and receptive principles. In like manner, the Principle of Gender is manifest in everything and not only in biological beings.

iv. Gender Complementarity: As a consequence of both the metaphysical derivation of gender and an equivalent physics-based derivation, the existence of complementarity between the genders can be derived as well. This is due to how the gendered units are individuated. In Mental Gender for instance we saw how the feminine "Me" was individuated from the masculine "I." The "Me" was cut from the "I," and was thus complementary to it. This can better be seen if we look at a more common example.

Imagine for instance a child cutting a star out of a piece of paper. Once the star is cut out, there remains the star and the paper it was cut from. Like the "I" in the previous example, the masculine principle would be representative of the paper, and the feminine would be representative of the star. Because the star was cut from the paper, the outlines of the star and the paper match each other perfectly. In other words they are "complementary."

Gender individuates in a similar manner, and thus gender complementarity is built directly into the Principle of Gender. Of course this would not manifest as literal cut-outs in reality. In the physical plane, men and women still appear due to biological reproduction.

The analogy is still valid though. In the mental and spiritual planes, the conscious states behind these physical states do still individuate from larger wholes. They may not have physical silhouettes, but the states contained in them match to each other in much the way that parts of a fragmented program fit together, resulting in complementarity.

B. Gender In Comparative Religion

A principle as important as Gender should show up elsewhere as well. If we do a little exploration, we quickly find that it can be found almost ubiquitously in comparative religion. This includes modern mainstream religions as well as ancient ones. A short summary of these can be found here.

1. Judaism: Perhaps containing the most popular example among Abrahamic religions, this has Eve being created out of Adam from Adam's rib. Regardless of how one interprets this, the archetypal significance is clear. Eve is cut out of Adam. It is further explained that this joint origin as a unified whole is the reason that "the two shall become one flesh."[2]

2. Christianity: In addition to sharing the example from the account of the creation of man, Christianity has another possible example to draw from. This shows up as a potential illustration of Mental Gender applied in the Divine case in the first chapter of the book of Colossians. Here all things are said to "hold together" in Christ, who is also described in the immediate context as the "head of the church."[3] Since Christ is the husband and the church is the wife, Mental Gender follows directly from this if Mentalism is applied.

3. Islam: The Sufi sect of Islam has a saying which reproduces the metaphysics of Gender as well. It says "Out of the original unity of being there is fragmentation and dispersal of beings. The last stage being the splitting of one soul into two. And consequently, love is the search by the other half for the other half on earth or in heaven."[4]

4. Buddhism: Tara and Avalokitesvara are bodhisattvas in Buddhism who are considered to be the female and male counterparts of one another respectively. According to legend, Tara came into being from a tear of Avalokitesvara being dropped into a lake.[5] Here we see the feminine principle originating from the male principle from a prior unification of the two in Avalokitesvara.

5. Hinduism: Shiva and Shakti are a god and goddess who represent the male and female principles respectively. They are considered counterparts of one another and are actually different aspects of the androgynous Ardhanarishvara who is considered to be the union of the two.[6] Again the pattern of male and female deriving from a unified whole appears.

6. Egyptian Religion: The god and goddess Osiris and Isis are considered to be male and female counterparts of one another, but are also brother and sister. This is not meant to signify incest. Rather it has cosmogonal significance as a metaphor in describing a common origin from a single whole.[7] Again male and female originating as a result of a split from a single whole is seen.

7. Sumerian Religion: In the cosmogony of the ancient Sumerians the world was created from the mountain Anki. This creation was caused by Anki splitting into An or heaven, and Ki or earth. Additionally, An was male and Ki was female.[8] It should be noted that this is the oldest known religion outside of prehistoric animism.

The universality of this principle is clearly reflected throughout religious history as well. This is important for two reasons. Firstly, it demonstrates that the Principle of Gender is not merely something thought up by Hermeticism, with no external basis. This is relevant due to the fact that the principle is difficult to establish on empirical grounds alone. Similarly, and secondly, it hints again at a prisca theologia, thereby suggesting that the religious ideas concerning gender are not merely artifacts of human imagination, but seem to be describing something about a common underlying reality.

C. Hieros Gamos
i. Gender and Marriage: The term "Hieros Gamos" is Greek for "holy marriage," and is the basis for the practice of marriage as understood in terms of the Hermetic Principle of Gender. Traditionally, it was described as the union of a god and goddess as represented in the form of a man and woman.

What this is representing is something deeper than just the marriage of god and goddess. It is an illustration that the marriage between masculine and feminine is more than the surface representations of male and female bodies in the physical plane. As we learned in the second and third chapters, the physical world is a mental construct produced by mental and/or spiritual planes which lie beneath it.

The physical world is an illusion, but the symbols in it are not random or meaningless. As we learned before, the physical world can quite accurately be thought of as a "natural simulation." Taking this description further, it can be seen that only a male or female "source code" would correspond to the "symbol" of a male or female body in the physical plane. The bodies are not only biological beings as we might think by limiting our view to the physical plane alone.

Hieros Gamos therefore is indicating that a deeper spiritual unification between male and female principles has been achieved through the union of these two bodies. This deeper unification is possible due to the fact that the genders are complementary to each other, just as the star and paper were as described before.

ii. Gender and The Fallen State: If we recognize that mental and spiritual planes also constitute the inner world; another interesting parallel emerges. The apocryphal Gospel of Thomas attributes Jesus to describing something which sounds eerily similar to Hieros Gamos in these words:

"When you make the two into one, and when you make the inner as the outer, and the upper as the lower, and when you make male and female into a single one, so that the male shall not be male, and the female shall not be female: . . . then you will enter [the kingdom]."[9]

The age and validity of the *Gospel of Thomas* are disputed and so these may not be Jesus' actual words. Nevertheless, whether it was influenced by later Neoplatonic thinking or these were Jesus' actual words, this saying precisely describes the Hermetic understanding of the unification of the genders. On a Hermetic understanding of Hieros Gamos, male and female becoming one in a manner that the inner and outer worlds reflect each other describes well how the masculine and feminine principles merge on both planes.

The reference to entering the kingdom is also significant as it describes an exit from the fallen state. In the fourth chapter we examined the idea of the fallen state as it shows up in comparative religion, and how this is paralleled by an examination of the topic in light of Mentalism and Correspondence. Here we discovered that the fallen state constitutes fractured elements of consciousness in the mental and spiritual planes. What was not explained before is that these fragments possess genders! This should not be unsurprising given that they derive from previously unified wholes, and that gender arises from the individuation of these wholes.

The exit from the fallen state then signifies a different kind of "marriage," although still a marriage between gendered elements. This is the "marriage" between the unified state of The Mind of The All with the fractured contents within it. The "marriage" between God and man! This makes sense in light of Mental Gender as it is applied to The Mind of The All. The Mind of The All's "I" is masculine, and its "Me" or contents are feminine. What marriage is in the microcosm, the unification of these fractured states to The Mind of The All is in the macrocosm.

iii. Abuse of the Principle of Gender: The reunification of dissociated genders brings with it additional power, a power which innately belongs to humanity. The complementary aspects of the masculine and feminine principles work together to produce a power that neither possesses alone. This fact is not widely known, due to widespread ignorance of gender and of the working of the Hermetic principles in general.

This ignorance is exploited by those who benefit from humanity not having their own power. Thus by attacking this principle, those without knowledge of it can become weakened and dependent on those who do. This is done in the modern day through an assortment of means, movements which split the genders against one another either through action or reaction, movements to try to erase the nature of marriage, and more recently movements to create confusion around or even destroy the concept of gender altogether.

To those with an understanding of the Hermetic Principles the nature of these sorts of movements should be transparent. When the principles of Mentalism, Correspondence and Gender are understood, biological sex is recognized to be merely one example of gender rather than something separate and divorced from it. Likewise, that the genders have complementary aspects which work in a synergy can not be ignored by those with an illumined understanding. Denial of such things can only be the result of ignorance rooted in the foolish ramblings and theories of those whose thinking is constrained to the physical plane, and of the great unwashed masses who allow their thinking to be determined by the Principle of Cause and Effect, thereby allowing themselves to be led like sheep.

The reader now informed of these principles can better see what is taking place so as to be able to help take back his or her power. Such efforts are only propaganda by the elite in the know who benefit from the disempowerment of humanity. It is spread by means of the promotion of unwitting tools who serve as false authorities in their own right, but who are in reality merely fools. Such propaganda should simply be ignored, and the real truth behind these principles should be spread.

As the Principle of Gender is the seventh and final of the Hermetic Principles, this concludes our study of the Hermetic Principles. The reader has hopefully gained much insight from understanding these principles and that there is a scientific underpinning to them. What is not explained though is how such principles could have been known, much less known in the times that they were, given the advanced nature of this knowledge. This is the topic we will turn to in our final chapter.

IX Epilogue

As the reader has been shown throughout the book, a great many occult concepts are rooted in real science. Such a realization is difficult to conceive for the materialistic age in which we live. This was especially evident in the chapters on vibration and polarity, which replicate precise details of physics known only very recently. Not only that, but it became apparent that the Hermetic principles are such that they can be used in ways that are dismissed in our modern era as impossible or non-existent, and that there is a scientific and rational way to account for these phenomena.

As was also explained in *The Kybalion*, these teachings did not originate in 1908 either, but were passed down from the distant past and were occulted or made hidden from the public eye. Thus, not only did the ancients have access to this science, but they were also able to use it in ways hitherto unrecognized and dismissed by modern science. How it was possible for this knowledge to have been originated is truly a mystery, one that we will attempt to address in this final chapter.

A. The Origins of Hermetic Science

i. The Mystical Hypothesis: Hermetic tradition itself says that Hermes Trismegistus originated this knowledge. But the question of how this was possible still exists. Was Hermes some mystic who was able to gain this knowledge through mystical intuition alone? We were after all able to derive Mentalism and Correspondence from the direct facts of our conscious experience alone, despite the fact that they can both be formulated in a rigorous scientific framework. This would be reasonable were it not for the other facts we have just observed.

Firstly, as our studies regarding vibration and polarity indicated, the details concerning this principles neatly matched things that were found to be the case by modern science, sometimes very modern science. The science of the quantum wave was replicated in *The Kybalion* with an uncanny accuracy some eighteen years prior to its actual scientific discovery in 1926 alone, much less after this knowledge was first originated prior to its writing. Further facts concerning the relationship of these vibrations to mentality, to the fabric of space and time, and to the universe at large came quite some time after that, yet were also replicated in Hermetic teachings. It is hard to imagine how anyone, mystic or otherwise would have known that vibration is present at scales far beyond their perception prior to the advent of twentieth century science, much less that these vibrations have a whole array of unique properties associated only with the waves of modern physics.

Secondly, is the integration of this science with the physics formulations of Mentalism and Correspondence. How they would have known to integrate this knowledge in a physically precise fashion remains a mystery. This is especially important given that it would require a knowledge of Mentalism and Correspondence beyond the mere mystical aspects of it. That "vibrations" exist for instance and correlate to mental states so as to predict things like quantum cognition and quantum smell is peculiar indeed.

Thirdly, is the question of how so many seemingly practical occult applications were developed from this. If one were to start from mysticism alone it is difficult to see how or why they would have thought to develop so many applications of this knowledge, and in such a fashion that follows from the integration of science that they should not even yet have had access to. The manner in which these are treated does not seem indicative of mysticism either. Mysticism pertains largely to *experiences*, but these things are and have been treated in an exacting manner as *occult sciences*.

For all these reasons, it seems unlikely that such a highly developed system should arise in this fashion. Even more unlikely is that it should integrate scientific concepts that would not arise for perhaps centuries. Thus to discover its origins, we should trace them back as best we can.

ii. Ancient Egypt: As described in the first chapter, the origins of the existing Hermetic texts can be traced back to the time of the burning of the Library of Alexandria. It was also explained that these writings were likely based on an earlier tradition passed down from Egyptian mystery schools. Can we trace these concepts back to ancient Egypt?

The ancient Egyptians had knowledge that was clearly advanced for their time, with mysteries that still captivate the imagination of the public and are not adequately explained. For example, the construction and accuracy of the pyramids, the Osireion, and similar sites are difficult to explain and impossible with modern engineering techniques. Controversies have appeared regarding the age of some of these sites as well, which do not fit into the expectations of mainstream archeology. Given this level of advancement is it possible that the Egyptians originated these concepts?

This hypothesis fails for obvious reasons. The concepts involved would indicate a working knowledge of other related concepts in physics as well. Yet the ancient Egyptians new little about electricity for example, much less the science of quantum mechanics. Other sciences one would expect a civilization with such knowledge to possess, such as advanced chemistry, genetics and microbiology, are lacking in ancient Egyptian culture despite their remarkable level of advancement.

Perhaps a good parallel would be with our culture today. Our society is more advanced than that of ancient Egypt in many ways. Yet as is described in this book, knowledge of Hermeticism appears in our culture as well. But how do we take it? Despite our own level of advancement, it appears to be an enigma or an out of place artifact that simply does not belong. Similarly in ancient Egypt, despite all of its mystery, it appears as an out of place artifact there as well.

But if it does not belong here or there, where does it belong? It would naturally belong to a culture that would have had the comparable knowledge one would expect out of other fields had they had a parallel level of advancement. Thus it appears this knowledge is a *relic* of some earlier civilization and culture now lost to us. This *relic* then survived into ancient Egypt and later into modern times.

iii. The Antediluvian World: If we explore the hypothesis that this occult science is the surviving relic of some earlier time now lost to us, we find it has corroborating support in the mythology surrounding ancient Hermeticism itself. This particularly pertains to the mythology surrounding *The Emerald Tablet* and its origins.

According to legend, *The Emerald Tablet* was said to be contained in the Pillars of Hermes along with more than three hundred scrolls.[1] They were said to have been moved from there to Heliopolis by Alexander the Great. In his writings, the Egyptian historian Manetho indicates that their origins were in the antediluvian world:[2]

"After the Great Flood, the hieroglyphic texts written by Thoth were translated from the sacred language into Greek and deposited in books in the sanctuaries of Egyptian temples."

It is interesting to note that he dates their origins as being nine thousand years old.[2] Given that Manetho lived in the 3rd century BC, this would place the date of "the Great Flood" at around 9300 BC if we were to stick to Manetho's chronology. This interestingly also corresponds to the date Plato gives for a cataclysmic flood.[3] Given the global distribution of myths of a great deluge and of an antediluvian world[4] it is unlikely that these stories are not based on actual events.

It is perhaps no coincidence that this date corresponds to the end of the previous ice age.[5] It is also worth noting, that ruins indicative of previous cultural advancement have been found in places such as Gobekli Tepe, and date to immediately after the end of the Pleistocene at around this same period.[6] Given these parallels it is reasonable to conclude that some catastrophic global event occurred between the boundaries of the Pleistocene and Holocene around which these floods myths are based, which presumably involved a massive cataclysmic meltdown of the ice caps.

Possible evidence for such a civilization is scant due to the level of destruction that would have been involved in such a catastrophe. Though some evidence has been uncovered which would lend credence to such a hypothesis.[7] Additionally, researchers such as Rose and Rand Flem-Ath have come up with theories with some merit about the possibility of such a civilization having existed in portions of Antarctica that were presumably at the time ice free.[8] This may parallel accounts of the so-called "White Island" in the Hindu Puranas[9] and is also reflected in contemporary knowledge of the Piri Reis map depicting accurate sub-glacial topography of Antarctica.[10]

All of this can be taken as a plausible hypothesis, and indeed the writings of the historian Manetho give us some support for it, but it still does not answer the question. *When* Hermetic knowledge was originated is not the issue. *How* it was originated and *by whom* are the real issues.

Even with an antediluvian civilization there are problems present. A civilization capable of developing this knowledge would also need to be a civilization capable of developing quantum mechanics, and at least some aspects of quantum gravity such as string theory, among other sciences. To do this, a civilization would need to have a significant level of advancement *and its associated infrastructure*. Such a civilization would also develop things such as a space program, nuclear reactors, particle accelerators, a massive road system, as well as a widespread electric grid and plumbing system.

Simply put, it would not be possible to hide evidence for such a civilization, even if it was destroyed in a massive cataclysm. Out of place artifacts have been found which do seem to indicate a level of out-of-place advancement.[11,12,13] Despite this, things like nuclear reactors and massive electric grids are found nowhere in the archeological record on earth.

Science is unfortunately not always good at giving us answers. However, what it is very good at is *ruling out* possible answers to the questions we ask. In this case, the science of archeology seems to be ruling out where the answer *is not*. It is hard to say where the answer comes from. One place we can say it *is not* though, is where there are no buried twelve thousand years old ancient electric grids or no ruins of ancient particle accelerators. Given that none of these things can be found on earth, yet would stick out like a sore thumb, we can rule out one place for the origins of this science; earth.

iv. The Intervention Hypothesis: Given where the answer *does not* lie, this leaves us with a very uncomfortable elephant in the room. If the origins of this occult science do not lie on earth, then they must by process of elimination lie "beyond earth." In other words, they must be "extra-terrestrial." Of course any "respectable" or "intelligent" person would immediately shun such a conclusion like the plague, just as they would go on to congratulate the emperor on his fine clothing.

The reader should be more intelligent than to be dissuaded by such things. After all, he certainly does not wish to be a part of the herd whose thoughts are determined by others through the Principle of Cause and Effect! Nor should he be worried about holding to opinions that he can keep in the privacy of his own mind away from the eyes of his peers, unless he is too embarrassed to even be honest with himself.

When we get past these concerns though, we can free ourselves to investigate such a hypothesis in a serious fashion and with an open mind. If we do this, the next logical step is to examine ancient texts about the history of pre-flood world. When we examine such texts the answer pops out at us immediately.

The Jewish Book of Enoch describes the pre-flood world in great detail. Included in its writing is an account of a race of fallen angels called the Grigori or Watchers who came down to earth. Here it explicitly tells us that these fallen angels taught mankind these occult sciences among other sciences and technologies.[14]

These beings are paralleled in other mythologies as well. They are identified with the Sumerian Igigi, a race of lesser gods beneath the Anunnaki.[15] The Akkadians referred to them as the Apkallu, wherein they were explicitly identified with Watchers and assumed the same role as teachers,[16] and included an association with witchcraft.[17] The Egyptians also recounted them under the name Urshu, and were seen as a lesser race of gods under the Neteru.[18]

Any talk of these beings is automatically identified with mythology in mainstream circles. While this is to be expected, several obvious problems exist with this approach. If these beings are fictitious then the historical record should have reflected that. Pieces of fiction are easily identified in history. In our own culture for example when fiction is written it is not treated as factual. Even if it evolves over time people recognize its fictional origins, and if it were ever to be taken as fact there would a period of cultural transition where this change was recognized to have gradually occurred. People do not naturally suddenly believe in fiction for no reason.

If these accounts are fictitious, then the closer one gets to their origin the more one would expect the people of the time to recognize their fictitious nature and not take them as true. To draw parallels with the modern world, today we recognize movies like *The Godfather* or *Independence Day* as fictional. They would be *less likely,* not more, to be believed as true one thousand years from now.

What is seen in the case of these accounts is the exact opposite. The farther from their origins one goes, *the more* they are seen as fictional rather *than less*. The farther back in time we go however, *the more* we find that these accounts are believed. When we go back to the earliest records of them, we find that they are taken in an almost matter-of-fact historical fashion by the cultures that originated them. This trend of belief is the opposite of what one would expect if they originated as fictional accounts that everyone recognized as fictional.

The parallels in these accounts deserves mention as well. Fictional accounts are unlikely to repeat in diverse civilizations, especially civilizations prior to the advent of modern communication and transportation. Moreover, any cross-cultural transference of information would not likely result in the dominant beliefs of an entirely separate culture being changed as we see here.

Finally, there is the fact that these occult parallels exist as an out of place relic, revealing indications of advanced modern physics, coming from comparatively unadvanced cultures. This fact alone indicates some kind of intervention and is not well-explained by mere coincidence. The further fact that the cultural traditions surrounding this subject trace their origins back to accounts of deities in an antediluvian world corroborates this thesis.

At this point, we must remark that the nature of these beings is not particularly relevant. Whether they be aliens or fallen angels is not the point. In fact, as we noted in the fourth chapter, this distinction is largely non-existent anyway beyond the limited categories imposed upon them by religious thinkers on the one hand, and materialistic thinkers on the other. What is important is that such an intervention took place, and how it has impacted and continues to impact the world even to this day.

B. The Abuse Of Occult Science

In this book the reader was introduced to many occult phenomena and shown the underlying principles at work behind them. Further it was shown that the occult science behind them has a solid basis in modern scientific principles. Sometimes the parallels with the science was very explicit.

Given the reality of these occult sciences, one may ask why they are hidden, or *occulted*, as they are. The second question this raises is what might the consequences of such knowledge being hidden be? What would be gained by concealing such knowledge, and how would those who do conceal it benefit from its concealment? By answering the second question first, the answer to the first becomes apparent.

i. The Influence of Occulted Knowledge: The answer to this question is obvious when one thinks about it. Knowledge is power. This means that if some knowledge is hidden, the power it enables its owners is power that those who do not know do not possess. The disproportion of knowledge gained by hiding or occulting some knowledge then translates into a disproportion of power. This disproportion of power between the few and the many is called *leverage* and gives the few an unfair advantage over the many. The knowledge of occult science is especially powerful and deals with the deepest forces in nature, forces which due to their profound nature are even dismissed as a possibility by the ignorant many.
If the degree of leverage is proportional to the power differential, then the power differential produced by concealing this knowledge must be vast. This being the case, one would expect those who wield such a power differential to be in a position of inordinate power.

When this is examined, it is again found to be the case. Evidence of the rich and powerful being connected to occult practices can be found online by even the most casual of researchers. In recent years these rumors have been confirmed to be factual, despite being dismissed as conspiracy theories for many years before. Authorities such as WikiLeaks have documented Satanic and occult practices among prominent political figures and celebrities.[19,20,21] This evidence is a matter of open record, and is readily accessible on the public domain.[22]

Without prior knowledge of the reality behind this occult science one may be prone to dismiss such things as superstitions among the elite. Given what was demonstrated in this book though it would be unwise to do this or to dismiss the influence of occult power in general. Likewise, it would be unwise to underestimate those who wield such power behind closed doors.

ii. Why We Are Here: This also answers how this situation arose, as the motives for concealing this knowledge are obvious. Those with power, especially those with darker motives, do not want to give up their power, and will fight to conceal it. This may also explain why topics related to the history of occultism have seemingly been occulted as well or are otherwise not part of official record.

Certain aspects pertaining to both archeology and anthropology have been ignored despite there being documented evidence of these particular matters. This is not to say that these fields are disreputable in their own right, and the general conclusions reached in these fields on most other matters are correct most of the time.

Where the relevant evidence lies though is largely along the borders of areas that are not focused on in these fields. Anomalous discoveries are often made not so much contradicting the big picture but going beyond it into a bigger picture. These are often ignored and explained away.

A good example of this includes the Paracas skulls. The official story behind these obscure but noteworthy skulls is that they are the product of skull-binding. This story is demonstrably wrong though and the evidence refuting it readily available to the public. Skull-binding only deforms the shape of the skull but is incapable of increasing their volume, as seen in these skulls.[23] Also of note, is the uniform absence of a sagittal suture seen in these skulls but nowhere else in the human species.[24]

If this is not enough, their genetic tests give the final word on the matter. Tests of their mitochondrial DNA demonstrate that they are distinct from normal *Homo sapiens*.[25,26] DNA tests are incapable of lying, but most academics as we will soon see, are very easily manipulated. Due to possessing an overly feminized psyche, they are capable of having their beliefs easily controlled through Mental Gender in a manner not much different from the rest of the great throng.

iii. How We Got Here: If one looks into the reasons for this state of affairs, one finds why such topics are shunned and documented facts are ignored. The example of the Paracas skulls is not accidental. If we look further, we find that members of ancient royalty have similar skull structures.[27,28] These lineages continue to the present day, although in highly diluted form, and are even present among today's modern elite.

Tracing this back. We find that this derived from the god-kings who according to legend were descended from the same kinds of beings said to be responsible for the origination of this knowledge. The knowledge and power resulting from it was always kept from the masses by those in power. The resulting power structure they left behind continues to this day.

This is perhaps why these beings were seen as evil in the Book of Enoch. Such power was abused in this manner, resulting in terrible consequences for the majority of humanity. If these structures remain as they are they could be abused again, as they are already being abused on lesser but still serious scales. To stop this, this knowledge needs to become widespread so that people can be aware of it and guard against its influences. Unfortunately, such knowledge is impossible in an age dominated by materialistic thinking. This period of time is what was described as the Kali Yuga and also corresponds to the Mayan baktun governed by the "Triumph of Materialism" as seen in the eighth chapter. This period has recently ended however, and this fact provides a unique opportunity for the future. If we are wise, we can exploit this opportunity.

C. The Importance of This Knowledge to Humanity's Future

To correct this situation this power differential needs to be balanced. This requires large-scale education on these topics. This is easier said than done, however. Even with widescale awareness of this knowledge, many will have a difficult time accepting it. There are several reasons for this, but they ultimately boil down to irrational structures programmed into the psyche, oftentimes by means of those with the means of mass influence. To examine these, it is first necessary to examine the sorts of reactions common among people who will read this book.

i. The Expected Reactions: If we have done our jobs properly and if the reader is sufficiently open-minded, he or she will hopefully have recognized the many fascinating parallels between discoveries in modern science and occult phenomena. Our hopes in writing this book is that this will give our readership a newfound way of looking at the world around them so that they will not be blind to these influences. What should not be done though is to abuse this knowledge via occult means.

All this being said, we also realize that not everyone will take this book in the same way. Different readers will come away after examining the same evidence in entirely different ways. Generally, we expect people to react to the contents of this book in one of three primary ways.

1. *Open Acceptance:* Some will see this and readily adopt it. Not only will they adopt it, but they will not have any problem with talking about in public. Though such people are often seen as fringe, gradual

trends are increasingly changing that.

    2. *Private Acceptance:* Others will see this and readily adopt it, as well seeing the logic in it. But unlike those in the first group, they will be more hesitant about being open about it. This is understandable given the ordinary social pressures placed on people, and that people need to function within a normal social environment. We do not discourage this. In fact, just the opposite is true. People in this category have an unique opportunity.

    3. *Complete Rejection:* Others will see this and filter out all of the evidence that does not fit their preconceived notions. It is possible of course for some to disagree with aspects of our thesis but not all of it entirely, but that is not primarily what we are talking about. People in this category are usually, though not always, controlled by patterns of thought that will force them into a frame of mind that prevents them from being capable of seeing anything outside of the picture given to them.

  These three groups of people present challenges and opportunities respectively. It is dependent upon people in the first two groups to convince people in the third group. To do this, we need to examine why people are psychologically resistant to these concepts.
  ii. Structures of Psychological Resistance: If we examine why people reject this information, the primary reason given will likely that they do not think it is *accepted* knowledge. This begs the question though. Accepted by *whom*? Invariably when asked about this, they will defer to some *authority*. This is despite the fact that the sufficiently educated lay person is capable of forming a valid opinion on these matters based on science alone, independent of any authority figure who may claim to *speak for* science. Knowledge is based on proof not acceptance by others, and people *should* know this. That they do not, tells us something about their psychology. Unfortunately, this kind of thinking even affects those who would seemingly possess scientific authority of their own. This very much regards Mental Gender, as we will get to shortly. But first we will examine the specific psychological roadblocks responsible for this in the various fields.

    1. *Physics:* In physics the actual science very firmly supports the general thesis. The only thing commonly excluded in this area is the belief that mentality is somehow involved with fundamental physics. The reasons behind this are superficial and not connected to actual science however.
  If the standard Copenhagen interpretation is taken to its logical conclusion, then consciousness is very clearly involved. Most physicists do not actually believe their own science though.[29] Rather quantum mechanics is treated instrumentally as a tool to produce results and not as though it reflects actual reality. Additionally, many "interpretations" of the theory are given to avoid its conclusions when taken at face value. If this were done in any other field, it would be treated as pseudoscience.
  Similarly, the science of emergent space-time directly contradicts the doctrines of physicalism and materialism. The physical plane can not fundamentally exist nor be the only thing in existence if it is also emergent from something other than itself. This fact is likewise ignored on grounds of an instrumental treatment of science as something other than a means to obtain knowledge about reality.
  Lastly, the fact that information is a mental entity and yet is also what space and time emerge from clearly spells out the Principle of Mentalism. This is ignored on similar grounds though, with only a few stopping to think about the metaphysical implications of such a profound set of statements about reality. The science of mind is instead deferred to neuroscience with little thought or question.

2. *Neuroscience:* Materialism is assumed from the outset, as the common belief is that the only alternative is a religious dualism. This materialism is then equated with an extremely crude scientism about the study of the brain. Here the activities of the brain are equated with mental states in a ham-fisted manner based on the correlations between them alone *with no or little regard for the actual facts about the experience of consciousness*, or how these facts relate to the science of the brain on the whole. The authors hope the reader recognizes the specific fallacy here as that of *post hoc ergo propter hoc.*

Though neuroscience is a very useful field of research in its own right, it is fundamentally restricted by its third person methodologies, which can tell us nothing about a topic that involves first person facts by its very nature. This involves the additional fallacy of the category error. Needless to say, if the presumed authority of an entire field over a specific topic is systematically based on fallacies regarding the application of its methodology that authority is not an authority at all.

3. *Philosophy of Mind:* To its credit, philosophy of mind is more free of instrumentalistic and scientistic thinking than the previous two fields. Nevertheless, it contains large pockets of resistance in regards to materialistic thinking. The flaws in this are easily apparent to anyone who has examined the field, with some in the field even going so far as to deny the contents or existence of their own consciousness![30,31] The errors in the arguments for conclusions like this should be equally apparent to the average person who knows the facts of his or her own consciousness intrinsically.

The reasons for this are due to a deference to the "authority" of neuroscience over consciousness. As previously explained however, this "authority" is based on nothing more than the fallacies intrinsic to scientism; the category error and the post hoc ergo propter hoc fallacies.

4. *Archeology & Anthropology:* These are perhaps the least influenced in the list. They are largely correct in their own fields of study. Where they err though, is more of an error of omission than one of commission. Anomalies that do not fit the preferred picture are oftentimes ignored. Though these anomalies are usually only limited to specific topics within these fields. Significant exceptions to this do occur though where "ordinary" explanations are sometimes systematically force-fitted into extraordinary finds even when it is clear they do not fit.

The reasons for this are rooted in a subconscious bias against the mundane, as well as the overarching assumption of uniform progress. "Everyone knows" that we progressed from stone agers to enlightened modern man. Thus narratives which suggest that there may be lost episodes in human history or outside interventions are rejected. This is perhaps due further to a subconscious pride in the notion of modern man being the pinnacle of existence. In a similar fashion, "extraordinary" possibilities that might also indicate we are not the pinnacle of existence psychologically agitate such pride.

The unsaid problem at the root of all of these though is that of the equation of authority, in this case academic authority, with knowledge. As was pointed out above, a great many of those who will reject the evidence in this book are basing their rejection upon their attitude of *receiving* knowledge from one they deem to be an *authority*.

Authority is associated often with initiation and confidence, both masculine traits. Whereas, as was explained in Mental Gender, reception is a feminine trait. Reception is not bad, nor are feminine traits bad. However, any traits that are out of their natural balance are bad. Being told what to think without question and without willingness to consider other possibilities is an example of just such an imbalance. It is difficult to correct this imbalance given its widespread nature. What can be done in the meantime though is to use this fact against itself. Given the psychological inclination of many people to act as followers, this can be done on a widespread scale.

iii. Breaking Down the Roadblocks: Once these facts are understood, the solution to the problem is rather simple. The readers described in the first category before are very necessary in getting the word out. However those in the second category who accept this knowledge quietly are key to making the real change.

The authority that those who reject the evidence provided here defer to is, as was just explained, fundamentally flawed. What is needed to remedy this is to undermine this false authority and reclaim it rather than oppose it directly. This is where the second category of readers are of vital importance.

One can not claim authority by simultaneously claiming to challenge it. This is why people are more likely to trust discredited mainstream news networks than they are to trust those who identify as "conspiracy theorists" or those who are "challenging the system" or even "brave scientists who are pushing a new paradigm against the mainstream."

All language like this must be removed if the false authorities are to be removed. This is because this language brings with it the sense that the information being given is not "mainstream" or "accepted" or "established" in the first place. To the open-mind it should not matter, as all such considerations of authority when it comes to knowledge should be irrelevant. But as we discussed, Mental Gender is at play, and much of the masses allow their beliefs to be determined for them rather than taking responsibility for their own power. It is also hard to convince them to take it.

The second group of readers not publicly admitting to accepting this information, but privately acknowledging its veracity, have a golden opportunity here. They can simply act to take the purely psychological authority that is used to keep others from realizing the truth for themselves, and use it to instead promote the truth.

Of course this can not be done directly, as any such discussion of these conclusions is currently outside the bounds of *current* normalcy. However, this normalcy can easily be shifted by bringing to the fore topics *that are within current normalcy*, yet that lead to the same conclusions here.

For example, discussions of digital physics and the illusory nature of space and time are within the bounds of current normalcy. So too, if presented appropriately, are discussions of the immaterial nature of the mind, and the need for it to be integrated with fundamental physics. Once these are pushed into the fore, it then becomes possible to introduce people to the logical conclusions that follow from them, before they realize exactly what those conclusions are.

In this manner counterpropaganda can be used to quickly break down the mental strongholds keeping this from becoming mainstream. If this is done from all directions at once, and from within each field and to the general public all at once, these mental blocks and the false authorities supporting them will quickly crumble as if being dissolved in acid.

This can be done in any profession or walk of life and by anyone who unbeknownst to others accepts the conclusions. The choice is yours and yours alone. We hope you choose wisely.

References

Chapter I

1. Daniel Stolzenberg, Egyptian Oedipus: Athanasius Kircher and the Secrets of Antiquity, University Of Chicago Press, pp. 238-240 (2013)
2. Thomas McAllister Scott, Egyptian Elements in Hermetic Literature, (April 18, 1987)
3. Tom Campbell, My Big TOE, Lightning Strike books; 1st edition (December 9, 2007)

Chapter II

1. Max Planck, *Das Wesen der Materie* [*The Nature of Matter*], *Archiv zur Geschichte der Max-Planck-Gesellschaft*, Abt. Va, Rep. 11 Planck, Nr. 1797 (1944)
2. Wojciech Hubert Zurek, *Decoherence, Einselection, And The Quantum Origins of The Classical* p. 4 (June 19, 2003)
3. A. Einstein, B. Podolsky, and N. Rosen, *Can Quantum-Mechanical Description of Physical Reality Be Considered Complete?* Phys. Rev. 47, 777 (May 15, 1935)
4. Alain Aspect, Philippe Grangier, and Gérard Roger, *Experimental Tests of Realistic Local Theories via Bell's Theorem* Phys. Rev. Lett. 47, 460 (August 17, 1981)
5. Anton Zeilinger et al, *An experimental test of non-local realism*, Nature, volume 446, pages 871–875 (April 19, 2007)
6. M.E. Goggin et al, *Violation of the Leggett–Garg inequality with weak measurements of photons*, PNAS (January 10, 2011)
7. Anton Zeilinger et al, *Quantum erasure with causally disconnected choice*, PNAS 110(4): 1221–1226, (January 22, 2013)
8. Anton Zeilinger et al, *Experimental non-classicality of an indivisible quantum system*, Nature volume 474, pages 490–493 (June 23, 2011)
9. Tobias Denkmayr et al, *Observation of a quantum Cheshire Cat in a matter-wave interferometer experiment*, Nature Communications volume 5, Article number: 4492 (July 29, 2014)
10. Werner Heisenberg, *Physics and Philosophy*, (New York: Harper and Row, 1962), p.145.
11. Sean Carroll, *Against Space*, preposterousuniverse.com (November 10, 2010)
12. Fotini Markopoulou, *Space does not exist, so time can*, arXiv:0909.1861 (September 10, 2009)
13. Sean Carroll, *Does Space-time Emerge From Quantum Information?* preposterousuniverse.com (May 5, 2015)
14. Sean M. Carroll et al, *Space from Hilbert Space: Recovering Geometry from Bulk Entanglement*, arXiv:1606.08444 (June 27, 2016)
15. Juan Martin Maldacena, "The Large N Limit of Superconformal Field Theories and Supergravity". Adv. Theor. Math. Phys. 2 (2): 231–252 (January 22, 1998)
16. Niayesh Afshordi et al, *From Planck Data to Planck Era: Observational Tests of Holographic Cosmology.* Physical Review Letters, 2017; 118 (4) DOI: 10.1103/PhysRevLett.118.041301
17. J.D. Bekenstein, *Black holes and the second law*, Nuovo Cim. Lett. 4 737-740F (1972)
18. Juan Maldacena & Leonard Susskind, *Cool horizons for entangled black holes*, arXiv:1306.0533 (July 11, 2013)

19. Sean M. Carroll et al, Space from Hilbert Space: Recovering Geometry from Bulk Entanglement, arXiv:1606.08444 (June 27, 2016)
20. "Metcam Oh," *Rosedale Makes Case For Holographic Universe, The Metaverse Tribune*, (August 22, 2011)
21. Brian Whitworth, *Physical World as a Virtual Reality*, arXiv:0801.0337 (January 5, 2008)
22. Nick Bostrom, *Are You Living In A Computer Simulation?, Philosophical Quarterly*, Vol. 53, No. 211, pp. 243-255. (2003)
23. J.A. Wheeler Jr. *"Information, Physics, Quantum: The Search for Links" Proceedings III International Symposium on Foundations of Quantum Mechanics.* Tokyo: pp. 354-358 (1989)
24. Searle, John. R. *Minds, brains, and programs, Behavioral and Brain Sciences* 3 (3): 417-457 (1980)
25. Donald D. Hoffman & Chetan Prakash, *Objects of Consciousness, Frontiers in Psychology*, Volume 6, Article 557 p. 13 (June 17, 2014)
26. Daniel Toker, *Information Idealism and the Integrated Information Theory of Consciousness*, (October 2013)
27. Giulio Tononi, Consciousness as Integrated Information: A Provisional Manifesto, Footnote 14 (2008)
28. Max Tegmark, Consciousness as a State of Matter, Chaos, Solitons & Fractals Section D. p. 13ff (March 17, 2015)
29. Kobi Kremnizer, & Andre Ranchin, Integrated Information-induced quantum collapse p. 3 (May 5, 2014)
30. Jerome R. Busemeyer & Zheng Wang, *What Is Quantum Cognition, and How Is It Applied to Psychology? Current Directions in Psychological Science,* (June 10, 2015)
31. Peter D. Bruza, Zheng Wang, & Jerome R. Busemeyer, *Quantum cognition: a new theoretical approach to psychology, Trends in Cognitive Science*, Volume 19, Issue 7, p. 383–393, (July 2015)
32. *Hamish G. Hiscock et al, The quantum needle of the avian magnetic compass, PNAS, 113 (17) 4634-4639 (April 26, 2016)*
33. Gregory S. Engel et al, *Evidence for wavelike energy transfer through quantum coherence in photosynthetic systems, Nature*, volume 446, pages 782–786 (12 April 2007)
34. Jacob Aron, *Quantum theory of smell causes a new stink, New Scientist*, (January 28, 2013)
35. Mareike Gutschner, *Discovery of quantum vibrations in 'microtubules' corroborates theory of consciousness, Elsevier*, (January 16, 2014)
36. E. Roy John, *The neurophysics of consciousness, Brain Research Reviews,* Volume 39, Issue 1, Pages 1-28 (June 2002)
37. F. Beck and J. C. Eccles, *Quantum aspects of brain activity and the role of consciousness, PNAS* 89 (23) 11357-11361; (December 1, 1992)
38. Matthew P. A. Fisher, *Quantum Cognition: The possibility of processing with nuclear spins in the brain*, arXiv:1508.05929 (August 29, 2015)
39. Jennifer Oullette, *A New Spin on the Quantum Brain, Quantum Magazine*, (November 2, 2016)
40. Tom Campbell, *My Big TOE, Lightning Strike books*; 1st edition (December 9, 2007)
41. Donald D. Hoffman & Chetan Prakash, Objects of Consciousness, Frontiers in Psychology, Volume 6, Article 557 pp. 11-13 (June 17, 2014)
42. Ekaterina Moreva et al, *Time from quantum entanglement: an experimental illustration*, arXiv:1310.4691 (October 17, 2013)

Chapter III

1. Sean M. Carroll et al, Space from Hilbert Space: Recovering Geometry from Bulk Entanglement, arXiv:1606.08444 (June 27, 2016)
2. Peter D. Bruza, Zheng Wang, & Jerome R. Busemeyer, Quantum cognition: a new theoretical approach to psychology, Trends in Cognitive Science, Volume 19, Issue 7, p. 383–393, (July 2015)
3. George Musser, *Is Dark Matter a Glimpse of a Deeper Level of Reality? Scientific American*, (June 11, 2012)
4. C.G. Jung, *The Structure of the Unconscious*, (1916)
5. C.G. Jung, *Jung's Collected Works Vol.* 8 Chapter 13, *Princeton University Press*; 2 edition, edited and translated by Gerhard Adler and R.F.C. Hull (March 1, 2014)
6. C.G. Jung & Wolfgang Pauli, *Atom and Archetype: The Pauli/Jung Letters, 1932-1958*, edited by C.A. Meier, translated by David Roscoe (2001)
7. Reinhard Blutner et al, *Two qubits for C.D. Jung's theory of personality, Cognitive Systems Research II* 243-259 (2010)
8. C.G. Jung, *Memories, Dreams, Reflections* (1962)
9. W.W. Wescott, *Sepher Yetzirah or The Book of Creation* (1887) Chapter II 4-5
10. Michael A. *Nielsen & Isaac L. Chuang, Quantum Computation and Quantum Information. Cambridge University Press.* ISBN 978-0-521-63503-5 (2004)
11. https://www.quantiki.org/wiki/bloch-sphere
12. Ian Glendinning, *Rotations on the Bloch Sphere,* (May 20, 2010)

Chapter IV

1. Kenneth Ring & Sharon Cooper, *Near-Death and Out-of-Body Experiences In the Blind: A Study Of Apparently Eyeless Vision, Journal of Near Death Studies* (1997)
2. Pim van Lommel, *Consciousness Beyond Life*, (New York: HarperCollins, 2010) p. 21
3. ABC-CLIO, LLC originally, The Handbook of Near-Death Experiences, edited by Janice Miner Holden, et al., (2009) Table 9.1, p. 194
4. Eben Alexander, *Proof of Heaven: A Neurosurgeon's Journey into the Afterlife*, (October 23, 2012)
5. Peter Novak, *Division of the Self: Life After Death and the Binary Soul Doctrine, Journal of Near Death Studies* Volume 20, Issue 3, pp 143–189 (2002)
6. Peter Novak, *The Division of Consciousness: The Secret Afterlife of the Human Psyche* (1997)
7. Hermes Trismegistus, *The Divine Pymander*
8. Howard Storm, *My Descent Into Death: A Second Chance at Life* (2005)
9. Walter Semkiw, *Reincarnation Case of an Israeli Child who Recalled Being Killed with an Ax & Identifies his Past Life Murderer, Who Confesses, Institute for the Integration of Science, Intuition and Spirit, From Children Who Have Lived Before, by Trutz Hardo*
10. Dean Radin, *Entangled Minds* pp. (82-3) (2006)
11. J.B. Rhine, *Extrasensory Perception After Sixty Years* (1940)
12. J. B. Rhine and J. G. Pratt, *A review of the Pearce-Pratt distance series of ESP tests Journal of Parapsychology,* 18, 165—77. (1954)
13. Sangeetha Menon, Anindya Sinha, B. V. Sreekantan, Interdisciplinary Perspectives on Consciousness and the Self, Springer; 2014 edition (December 12, 2013) p. 237

14. J. Ambjorn, J. Jurkiewicz, R. Loll, *Emergence of a 4D World from Causal Quantum Gravity*, arXiv:hep-th/0404156 (September 16, 2004)
15. Norman Israel, Two Dimensional Causal Dynamical Triangulation, (Apr 29, 2011)
16. Amanda Gefter, *Concept of 'hypercosmic God' wins Templeton Prize, New Scientist,* (March 16 2009)
17. Peter Novak, *The Lost Secret of Death: Our Divided Souls and the Afterlife* Chapter 1, (2003)
18. Jerry A. Fodor, *The Language of Thought, Harvard University Press* (1975)

Chapter V

1. *What is the Schrodinger equation, and how is it used?* http://www.physlink.com/education/askexperts/ae329.cfm
2. *Free particle approach to the Schrodinger equation,* http://hyperphysics.phy-astr.gsu.edu/hbase/quantum/Schr2.html
3. Zeeya Merali, *The universe is a string-net liquid, New Scientist*, (March 15, 2007)
4. Clara Moskowitz, *Tangled Up in Space-time, Scientific American* (October 26, 2016)
5. *How space-time is built by quantum entanglement, Phys.org,* (May 27, 2015)
6. Michael A. Levin & Xiao-Gang Wen, *String-net condensation: A physical mechanism for topological phases,* (April 26, 2004)
7. Zeeya Merali, The universe is a string-net liquid, New Scientist, (March 15, 2007)
8. DeWitt, B. S., "Quantum Theory of Gravity. I. The Canonical Theory". *Phys. Rev.* 160 (5): 1113–1148. (1967)
9. Weighing the Entire Universe: Dark Matter and Dark Energy, https://imagine.gsfc.nasa.gov/features/satellites/archive/map_weighing.html
10. Schrödinger, E., "Quantisierung als Eigenwertproblem; von Erwin Schrödinger". *Annalen der Physik.* 384: 361–377. (1926)
11. Einstein, A; B Podolsky; N. Rosen "Can Quantum-Mechanical Description of Physical Reality be Considered Complete?" (PDF). *Physical Review.* 47 (10): 777–780. (May 15, 1935)
12. DeWitt, B. S., "*Quantum Theory of Gravity. I. The Canonical Theory*". *Phys. Rev.* 160 (5): 1113–1148. (1967)
13. Michael A. Levin & Xiao-Gang Wen, *String-net condensation: A physical mechanism for topological phases,* (April 26, 2004)
14. *How space-time is built by quantum entanglement, Phys.org,* (May 27, 2015)
15. Peter D. Bruza, Zheng Wang, & Jerome R. Busemeyer, Quantum cognition: a new theoretical approach to psychology, Trends in Cognitive Science, Volume 19, Issue 7, p. 383–393, (July 2015)

Chapter VI

1. Dean Radin, Gail Hayssen, Masuru Emoto, Takashige Kizu, *Double-Blind Test of the Effects of Distant Intention on Water Crystal Formation, EXPLORE*, Vol. 2 No. 5 (September/October 2006)
2. Jan Wicherink, *Souls of Distortion Awakening* p.83 (2007)
3. Belle Dumé, *Is the universe a dodecahedron? Physics World*, (October 8, 2003)
4. John Whitfield, *Universe could be football-shaped, Nature,* (October 9, 2003)
5. Michael Allen, *Birds measure magnetic fields using long-lived quantum coherence, Physics World* (April 7, 2016)

6. David Biello, *When It Comes to Photosynthesis, Plants Perform Quantum Computation, Scientific American* (April 13, 2007)
7. Schoenlein, R. W.; Peteanu, L. A.; Mathies, R. A.; Shank, C. V. *"The first step in vision: femtosecond isomerization of rhodopsin". Science.* 254 (5030): 412–415 (October 18, 1991)
8. *Quantum Entanglement Holds DNA Together, Say Physicists, MIT Technology Review,* (June 28, 2010)
9. Montagnier, L; Aïssa, J; Ferris, S; Montagnier, JL; Lavallée, C., *"Electromagnetic signals are produced by aqueous nanostructures derived from bacterial DNA sequences". Interdisciplinary Sciences, Computational Life Sciences.* 1 (2): 81–90 (2009)
10. Andy Coghlan, *Scorn over claim of teleported DNA, New Scientist* (January 12, 2011)
11. Charles Bergquist, *Caught On Video: How DNA Replicates, Science Friday,* (June 23, 2017)
12. Frank van den Bovenkamp, *"Why Phi" a derivation of the Golden Mean ratio based on heterodyne phase conjugation,* (March 5, 2007)
13. Cyril W. Smith, *Qi & Bio-Frequencies,*
14. Vincent Bridges and Teresa Burns, *Shakespeare And Dr. Dee, Atlantis Rising,* (May/June, 2009)
15. *Enochian Gene Keys,* http://deuterontherion.tumblr.com/post/124114131563/enochian-gene-keys

Chapter VII

1. Donald D. Hoffman & Chetan Prakash, *Objects of Consciousness, Frontiers in Psychology,* Volume 6, Article 557 p. 13 (June 17, 2014)
2. Donald D. Hoffman & Chetan Prakash, *Objects of Consciousness, Frontiers in Psychology,* Volume 6, Article 557 pp. 11-13 (June 17, 2014)
3. Giulio Tononi, *Consciousness as Integrated Information: A Provisional Manifesto,* Footnote 14 (2008)
4. Max Tegmark, *Consciousness as a State of Matter, Chaos, Solitons & Fractals* Section D. p. 13ff (March 17, 2015)
5. Kobi Kremnizer, & Andre Ranchin, *Integrated Information-induced quantum collapse* p. 3 (May 5, 2014)
6. Jacob Aron, *Quantum theory of smell causes a new stink, New Scientist,* (January 28, 2013)

Chapter VIII

1. William Drummond, *"Oedipus Judaicus - Allegory in the Old Testament", Bracken Books,* London, 1996, p xix, 159 (July 1996)
2. Robert Hackman, *Ancient Maya Holy Time and the Evolution of Creation Map,* Chapter 10 (December 2010)
3. John D. Smith, *The Mahābhārata: an abridged translation,* Penguin Classics p. 200 (2009)
4. Walter Scott, *Hermetica: The Ancient Greek and Latin Writings which contain Religious or Philosophical Teachings ascribed to Hermes Trismegistus* 1:341-7, translated from a now lost Latin text attributed to Apuleius (September 12, 1985)

Chapter IX

1. Roger Nelson, The Global Consciousness Project Meaningful Correlations in Random Data, Global Consciousness Project, http://noosphere.princeton.edu/
2. Roger Nelson, *Formal Analysis September 11 2001*, Global Consciousness Project, http://noosphere.princeton.edu/911formal.htm
3. Roger Nelson, *Formal Results: Testing the GCP Hypothesis, Global Consciousness Project,* http://noosphere.princeton.edu/results.html
4. Sebastian Anthony, *CERN now 99.999999999% sure it has found the Higgs boson, ExtremeTech*, (December 17, 2012)
5. Dean Radin et al, *Consciousness and the double-slit interference pattern: Six experiments, Physics Essays Publication 25, 2, DOI: 10.4006/0836-1398-25.2.157, (May 16, 2012)*
6. D. Wegner, *The Illusion of Conscious Will.* Cambridge, MA: MIT Press. (2002)
7. Anton Zeilinger et al, *An experimental test of non-local realism, Nature,* volume 446, pages 871–875 (April 19, 2007)
8. M.E. Goggin et al, *Violation of the Leggett-Garg inequality with weak measurements of photons,* arXiv.org (July 9, 2009)
9. Kim, Yoon-Ho; R. Yu; S.P. Kulik; Y.H. Shih; Marlan Scully "A Delayed "Choice" Quantum Eraser". *Physical Review Letters* 84: 1–5 (2000)
10. *Lady Gaga, The Illuminati Puppet, The Vigilant Citizen* (August 9, 2009)
11. *The Occult Meaning of Beyoncé's "Lemonade," The Vigilant Citizen,* (May 4, 2016)
12. *The Occult Semi-Subliminals of Jay-Z's "On to the Next One," The Vigilant Citizen,* (January 3, 2010)
13. *The Dark Occult Meaning of Nicki Minaj's "No Frauds," The Vigilant Citizen,* (April 25, 2017)
14. *Kanye West's "Yeezus": Surrounded with Occult Symbolism, The Vigilant Citizen,* (June 28, 2013)
15. *"Mexico_b" Lady Gaga: 'I Regret Selling My Soul to Illuminati Dark Forces,'* Anonymous.com (September 24, 2017)
16. *The Truth About Pepe The Frog And The Cult Of Kek,* https://pepethefrogfaith.wordpress.com/
17. Don Caldwell, *Shadilay,* http://knowyourmeme.com/memes/shadilay

Chapter X

1. M. Mills, A. Janiszewska, & L. Zabala, *Sex differences in making risky first time relationship initiatives* (2011)
2. The Bible, New International Version (NIV), Genesis 2:21-24
3. The Bible, New International Version (NIV), Colossians 1:17-19
4. Idries Shah, *The Sufis,* (July 7, 2015)
5. *Tara Tibetan Goddess of Compassion, Devi Press,* http://www.goddess.ws/tara.html
6. Aimee Hughes, *Shiva and Shakti, Yogapedia,* (July 30, 2016)
7. *Holy Matrimony – Hieros Gamos,* https://www.cs.utah.edu/~spiegel/kabbalah/jkm015.htm
8. David Adams Leeming, *Creation Myths of the World: An Encyclopedia, Volume 1, Part II: The Creation Myths* p. 247, (December 18, 2009)
9. *The Gospel of Thomas,* Translated by Stephen J. Patterson and James M. Robinson, Saying 22

Chapter XI

1. J.R. Sedivy, *The Emerald Tablet*, http://jrsedivy.com/the-emerald-tablet/, (October 7, 2016)
2. Dennis William Hauck, *The Emerald Tablet: Alchemy of Personal Transformation,* Chapter 2, *Alexander's Treasure, (*March 1, 1999)
3. Plato, *The Project Gutenberg EBook of Timaeus,* Translated by Benjamin Jowett, (September 15, 2008)
4. Alan Dundes, *The Flood Myth, University of California Press; 1st Edition,* (February 16, 1988)
5. Cesere Emiliani, *Earth and Planetary Science Letters*, 41 (1978), p.159, Elsevier Scientific Publishing Company, Amsterdam
6. "Oliver," *How old is it? Dating Göbekli Tepe., The Tepe Telegrams, (*June 22, 2016)
7. Karen Mutton, *Sunken realms: a complete catalog of underwater ruins, Adventures Unlimited Press*, (April 1, 2009)
8. Rose & Rand Flem-Ath, *Atlantis beneath the Ice: The Fate of the Lost Continent, Bear & Company; 2nd edition, (*February 10, 2012)
9. Unknown, *Vishnu Purana,* Book II, Chapters 1,2,&3. Translated by Horace Hayman Wilson Forgotten Books, (January 23, 2008)
10. Charles H. Hapgood, *Maps of the Ancient Sea Kings: Evidence of Advanced Civilization in the Ice Age,* Adventures Unlimited Press, (1997)
11. Nicholas Paphitis, *"Experts: Fragments an Ancient Computer," Washington Post*, (November 30, 2006)
12. A. Frood, *Riddle of 'Baghdad's batteries,' BBC News*, (February 27 2003)
13. P. Stromberg, and P. V. Heinrich, 2004, *The Coso Artifact Mystery from the Depths of Time?, Reports of the National Center for Science Education*, v. 24, no. 2, pp. 26–30, (March/April 2004)
14. Unknown, *The Book of Enoch,* Chapters VI, VII, & VII, Translated by R.H. Charles, (1917)
15. Fredsvenn, *The Anunaki, the Igigi and the humans, Cradle of Civilization,* (May 26, 2017)
16. Michael S. Heiser**,** *Reversing Hermon: Enoch, the Watchers, and the Forgotten Mission of Jesus Christ, Defender Publishing*, (March 24, 2017)
17. Amar Annus, *On the Origin of Watchers: A Comparative Study of the Antediluvian Wisdom in Mesopotamian and Jewish Traditions, Journal for the Study of the Pseudepigrapha*, 19(4):277-320, (May 2010)
18. Legends of Egypt, http://humanpast.net/legends/separate/egypt.htm
19. https://wikileaks.org/podesta-emails/emailid/15893
20. https://wikileaks.org/podesta-emails/emailid/16498
21. https://wikileaks.org/clinton-emails/emailid/14333
22. https://wikileaks.org/
23. Brien Foerster, *Elongated Skulls Of Paracas: A People And Their World, Hidden Inca Tours,* https://hiddenincatours.com/elongated-skulls-of-paracas-a-people-and-their-world/
24. *Skulls,* http://thegreaterpicture.com/skulls.html

25. April Holloway, *New DNA Testing on 2,000-Year-Old Elongated Paracas Skulls Changes Known History, Ancient Origins,* (July 23, 2016)
26. *DNA Analysis Of Paracas Elongated Skulls Released. The Results Prove They Were Not Human, Sunny Skyz,* (February 6, 2014)
27. John Nobel Wilford, *Long Skull, Narrow Face: Tut Gets New Look, The New York Times,* (May 11, 2005)
28. Tara MacIsaac, *A Look at Theories About Elongated Skulls in Ancient Peru, Europe, Egypt, The Epoch Times,* (July 30, 2014)
29. Eliezer_Yudkowsky, *Quantum Non-Realism, Less Wrong,* (May 7, 2008)
30. Reuben Westmaas, *There's No Such Thing as Consciousness, According to Philosopher Daniel Dennett, Curiosity,* (April 14, 2017)
31. Daniel C. Dennett, *Quining Qualia,* In: Marcel, A. & Bisiach, E. (eds.) *Consciousness in Modern Science,* Oxford University Press, (1988)

Printed in Great Britain
by Amazon